The Use of ICT in Higher Education

The Use of ICT in Higher Education

A Mirror of Europe

Editors
Marijk van der Wende
Maarten van de Ven

Lemma Publishers – Utrecht – 2003

ISBN 90 5931 181 7
NUR 841

http://www.lemma.nl
infodesk@lemma.nl

© 2003 LEMMA Publishers, P.O. Box 3320, 3502 GH UTRECHT, The Netherlands

Cover design and typography: Twin Design BV, Culemborg, The Netherlands

Contents

Preface

This book presents an overview of the state of the art with respect to the use of information and communication technology (ICT) in higher education in a wide range of European countries. The papers included in this volume were first presented at the Conference on the 'New Educational Benefits of ICT in Higher Education', which was held in Rotterdam in September 2002.

Besides the description and analysis of the use of ICT in eleven European countries (chapters 4-14), a number of thematical chapters are included. They address the role of European Union policy and programmes in this area (chapter 2) and their impact at the institutional level (chapter 3). The last two chapters highlight strategic issues of particular interest: the role of cooperation and competition (chapter 15) and the regional and cultural factors that interact with ICT in higher education (chapter 16). An introduction to the various chapters, including a general overview of the various themes and issues addressed is presented in chapter 1.

We would like to thank Stichting SURF (SURF Foundation) and Erasmus University Rotterdam, whose generous support has made publication of this book possible.

The editors,
Marijk van der Wende
Maarten van de Ven

1 A Mirror of Europe: The Use of ICT in Higher Education

Marijk van der Wende, CHEPS, University of Twente, The Netherlands
Maarten van de Ven, Erasmus University Rotterdam, The Netherlands

1.1 Introduction

Higher education is adapting to one of the most challenging developments in its history: the emergence of a society that is global, networked and in which knowledge is the main economic driving force. This development is at the same time the result of, as it is also facilitated by, the use of Information and Communication Technologies (ICTs). In their response, higher education institutions are challenged to integrate these technologies into their core processes and organisation, and to develop strategies for effectively educating their students for this new social context.

Who are these new students and what are their expectations? What role does ICT play in their current life and what role will it play in their future? And how can teaching and learning processes be adapted to these new learners and their needs and expectations? These are only a few of the questions that need to be answered in this respect. This book demonstrates the complexity of these issues. Over the last years, higher education institutions, national and European governments and agencies have tried to analyse and address the implications of this knowledge and technology-driven global society. The various chapters report on what has been achieved in these first years of web-based teaching and learning and on the lessons that can be learned from it.

1.2 Change Drivers, Policy Responses and Institutional Realities

It is generally recognised at policy level that demographics, globalisation, economic restructuring and information technology are the main forces driving change in the higher education sector. The resulting expansion and differentiation of enrolments, increasing competition and growing pressure for cost-effectiveness and efficiency enforce reform and innovation at both system and institutional level. Policy responses at national and supranational level concentrate on widening access and on improving quality and flexibility. ICT is seen as a major tool to both optimise the learning experience of the traditional on-campus students, as well as to enable the institution to reach out to new target groups in the lifelong learning and international markets.

Institutions acknowledge these aims as the main rationales for action. However, the various contributions to this book make it very clear that, despite the general understanding of these global trends, it is very difficult for institutions to respond fully and quickly enough to them. In many cases, there seems in fact to be a mismatch or a gap between the policy rhetoric about the knowledge economy and the more practical and incremental approach to ICT at institutional level. It is estimated that in Europe only few institutions have actually fully completed the process of strategic decision making, and that many institutional leaders feel ill-equipped to face these major changes. There is more often a general notion of keeping up with the competition and fear to be left behind in the ICT race, than an actual and clear vision on the role that ICT could and should play in the institution's mission and actions.

14

Governments generally take responsibility for the development of national network infrastructure, establish support agencies, and try to stimulate innovation and ICT implementation at institutional level. Collaboration is greatly encouraged as a mechanism that could ensure the clustering of investment and expertise, sufficient economy of scale and effective dissemination of results. Collaboration within the sector but also with corporate partners has led to a multitude of bottom-up networks and consortia, supplemented in several countries by more centralised initiatives leading to the establishment of national virtual universities. But as said before, and as described in the following country chapters, the link between these policies and the institutional reality is often problematic. First, there is the complex and sometimes inefficient interplay between the policy actors at various levels. Horizontal policy coordination between EU directorates and between national ministries often represents important challenges. Vertical coordination between European, national (regional) and institutional levels may also be quite complicated. Second, effective policy links are missing in the area of incentive and reward systems at the level of the institutions and for the individuals working in them. In universities career trajectories and professional status are primarily based upon research results. Therefore, ambitions of staff members are focussed upon conducting research rather than teaching. Third, there is a lack of feed-back on policy initiatives on the basis of systematic evaluation and real evidence. In many cases institutions are not very active in assessing and monitoring their ICT developments, which complicates adequate data gathering at more aggregate levels. These problematic policy links make it sometimes difficult to determine to what extent policies matter and leave questions open on what is strategy and what is serendipity.

This is not to say that initiatives at institutional level have been lacking. On the contrary, technical infrastructure has been put in place and rich experimentation with the educational use of ICT has been undertaken. The transformative role of ICT is being recognised and expressed in many institutional documents, but in most cases the strategic impact of ICT has yet to be realised. The large ranges of projects in which institutions have invested have stimulated an agenda of bottom-up innovation rather than one of major institutionally-led strategic initiatives. Lessons so far demonstrate that the implementation of ICT is an evolutionary process and that indeed few revolutions have happened so far. It is mainly a process of bottom-up, incremental change from within through which the use of new technologies is integrated into the old and existing practices. Blended models of teaching and learning occur and institutions that combine 'brick and click' seem to be most successful.

15

1.3 Choices and Scenarios for the Future

For the next phase, which would logically focus on the further institutionalisation and on the sustainability of these innovations, we need to know more about how these blended models actually work; and which combinations are most effective with respect to learning processes and outcomes and for which particular target groups. We also need to explore what the combination of concrete and digital infrastructure means in terms of organisational designs, processes and budgets.

Various authors in this book explore the possible answers to these questions and their implications in terms of choices and scenarios for the future. The different models that are presented vary in a number of ways from each other, but together they bring an interesting combination of perspectives to the fore. They vary first of all with respect to the extent to which ICT is integrated in the institution. A low profile approach implies the use of technology only for enhancing effectiveness, a medium profile the systematic integration of technology, and a high profile a true virtual university concept. Secondly, they vary with respect to the consequences that strategic choice in this area will have for the organisational structure and culture. Here the question is whether the university as an organisation will preserve its character of a professional bureaucracy, change itself into a divisional organisation or transform itself into a radically new type of university. A third dimension is the impact of ICT on the sector as a whole. Will provisions become de-institutionalised and convention-al education and training institutions lose importance to new providers? Will the traditional sector innovate enough and transform itself through an increased provision of services that help learners to become autonomous and by getting lifelong learning in place? Or will a 'dual market' develop in which degree-oriented learners will continue to address conventional institutions while most competence-oriented learners will find other suppliers who can better serve their requirements?

The combined effects of demographical change, international competition and of ICT itself will not leave the curriculum, its structure and certification mechanisms unchanged. Life-long learners and also international students may be more interested in taking individual courses than complete study programmes at a particular institution. They may combine these with courses taken elsewhere and earn a final award or degree through portfolio assess-ment. The questions are whether higher education institutions will be able to accommodate these new types of learners, how can they use ICT in doing so and how to combine these new arrangements with traditional campus-based teaching and learning.

1.4 Institutional Strategies

In general, institutional strategies on the use of ICT in education are based upon responses to external developments (e.g. technology push, national policies) and do not seem to rely upon explicit new educational visions. At institutional level most universities choose to use ICT to strengthen their traditional educational philosophies, which generally concentrate on campus-based courses, a face-to-face instructional approach and traditional students as the main target group. In many of the continental European countries, the new types of learners, as discussed above, are not (yet) seen as a major new target group by the institutions. In the competition that emerges in this new market, institutions sometimes seem to strengthen their traditional values as their main weapon against new providers of distance learning solutions, instead of looking at how they can broaden their own mission and services by using ICT.

But even though the educational philosophies remain mostly unchanged and implicit, the educational processes do change gradually. Initiatives for implementing learner-based philosophies or a mix of teacher-learner-based educational formats are undertaken at decentral levels. Almost all institutions have explicit strategies in implementing new technologies to improve the quality and efficiency of their primary and secondary educational processes. In the last couple of years these strategies were aimed at implementing digital learning environments, based upon course management systems and administrative software. The implementation of course management systems such as Blackboard or WebCT is aimed at improving primary educational processes, by offering digital resources, electronic communication and personal tools such as electronic diaries. Administrative software, for example used for student registration and grade distribution, is implemented for improving secondary processes.

In general, the main aim of the institutional implementation strategies lies in enrichment of lectures by means of ICT and in providing more flexibility for the traditional student. In lecture-based education the digital learning environment is used for distribution of documents and for communication. Examples of documents that are made available through the digital learning environments vary from lecture notes and sheets to video recordings of (parts of) the lectures. Examples of communications range from the use of e-mail to ask teachers questions before or after the lectures to the use of discussions boards on the content to discuss propositions in small groups and between groups. Furthermore, ICT is used to create flexibility at different levels. Within individual courses, ICT is used to offer educational activities to students and groups of students independent of time and place. Within study programmes, ICT is used to enlarge possible combinations of courses within one study programme. Through this enlarged flexibility in time, place and programme the students get more possibilities for 'virtual mobility'.

Next to these general aims, institutions have created facilities for specific educational innovations within individual courses or study programmes. In most institutions, funds are available to create new types of educational formats, which move away from teacher-led education to learner-led education or a combined form of teacher-learner-led education. A mix of new types of education can be distinguished, in which teaching functions are gradually taken over by the individual student or groups of students. Examples of these functions are defining course objectives, planning and controlling student activities, setting up procedures for assessment and presenting course material. Although a large number of small, innovative projects are performed, their impact on the general educational vision of institutes is only limited.

The attempts by institutions to reach these aims might lead to the following future developments.

- *Learning in groups as opposed to teacher-led education will gradually become a more dominant type of education.* Many of the learner-centred types of education are based upon group learning, either in small groups or teams or in larger communities. In these learning communities (e.g. Communities of Practice) different participants, such as teachers, learners or experts might have different roles within that community. Students will learn by working together in groups of students by working on projects, problems solving tasks, etc. Meanwhile, their teachers in a role as tutors or guides will coach these groups. Furthermore, experts in the field in question are available at a distance. ICT is used to facilitate the communication in all directions: within groups, between groups, between students and their teachers and between groups and experts. New communication patterns will be developed to structure these types of communication.
- *New ways of organising educational processes will lead to more distinct specialisation in educational processes.* On the one hand, the integration of ICT in education leads to more complex educational formats. On the other hand, the changing paradigm of teacher-led to learner-centred types of education demand a broader focus on student competencies. From the very start of the study, assessments will be aimed not only at cognitive objectives but also on affective, social and connotative objectives. In other words, the complete behaviour becomes important. This means that both the development of these new educational formats as well as the actual teaching processes themselves become more specialised. Teaching will become more a matter of teamwork, in which teachers, tutors, assessors, content experts, educational technologists and students work together. While, in the past, most of these different functions were all performed by teachers themselves, in the future, collaboration between specialists in these areas will become the main way of organising educational processes.

– *Content management is the next step in implementing digital learning environments.* Three reasons underlie this assumption. First, after having implemented digital learning environments, many institutions are now realising that a huge amount of content is available in this environment. To maintain this capital of educational resources at a level above that of the individual teacher, a content management system is needed. Second, such a system is required to control the life-cycle of these resources. If education is a matter of team-work of different specialists, workflow management of the development and use of educational formats and resources is very important. Third, a content management system can be used to control the delivery of resources in terms of calendar time and target groups. Institutions might use this feature to make tight agreements with publishers.

– *Interactive educational software will complete the more linear resources in the digital learning environment.* Between 1980 and 1990, huge effort was put into the development of interactive, educational software packages. This software failed to meet high expectations in terms of improving the efficiency and quality of the educational processes that were implied by their use. Furthermore, the development of these products was very expensive. This is partly due to the not-so-sophisticated authoring tools that were available at that time. Now, authoring tools are far more sophisticated. In combination with the digital learning environment most institutes have available, these tools might be used for cost-efficient development of interactive educational software.

The above, but especially also the following contributions to this book demonstrate the complexity of managing ICT in higher education. Institutions have at the same time to keep up with the newest technological possibilities and challenges, and to ensure smooth implementation in both primary and secondary processes. They have to develop new educational models that correspond to students' needs and that employ ICT to its full extent. But above all they have to define their (new) role in a global knowledge society. Strategic choices around ICT are fundamentally related to the mission of the institution regarding the different types of learners in this society, to its regional, national and/or international reach and to its position with respect to partners and competitors.

2 EU Policy in the Field of ICT in Higher Education

Maruja Gutiérrez Díaz, DG Education and Culture, European Commission

2.1 Education: a Priority Policy for the European Union

The Lisbon Summit, 23-24 March 2000, is a landmark for EU policy. Its conclusions have become the European brief for the first ten years of the new century. A lucid and courageous analysis of the problems and opportunities facing Europe in the passage to a new era, and a review in detail of the main social, economic and cultural challenges ahead, the conclusions call for concerted action with an ambitious vision: to become the most competitive and dynamic knowledge-based economy in the world, capable of sustainable economic growth with more and better jobs and greater social cohesion.

The impact of the Lisbon Conclusions has been particularly strong for education policy. For the first time, education becomes a top rank EU policy, indivisible from economic and social policy. In a knowledge-based economy, education is a key asset. One of the first actions required is the adaptation of education and training systems.

This need for change and adaptation is not a discovery from Lisbon; practically all Member States are engaged in a process of revision of education systems, to provide an adequate answer to Information Society needs. However, Lisbon represents an important change in three aspects.

The first and main one is the clear assertion of *education as a common concern*, with a European dimension. Education is, as culture, a 'subsidiarity' sector i.e. one where competences remain exclusively with Member States. Yet, there are advantages to be gained in setting common objectives for the whole of the Union, in joint monitoring of their achievement, and in increased information and exchange of experienceusing agreed methods, such as benchmarking or peer-reviewing.

The second is the progress from an 'Information Society' approach, with a sometimes overpowering consideration of technological aspects, to a 'Knowledge Europe'. *The emphasis is clearly placed on the human factor.* Education, training, and research are the essential grounding for innovation and competitiveness, the engines of sustained economic development, with more and better jobs, and improved social cohesion.

The third is the embracing of *lifelong learning as overall educational paradigm.* This change encompasses a global change of approach, fostering a 'culture of learning', for education, work and personal fulfilment. Addressing the vast range of new education and training needs coming from this change of paradigm will need a large re-thinking of education systems.

2.2 An Overview of Recent EU Actions for ICT in Education

The European Commission has led pioneering actions in this field for many years. First, under the research programmes (ESPRIT, and in particular DELTA, the first action dedicated specifically to learning technologies); then under the education and training programmes *Socrates* and *Leonardo da Vinci*. In 1997, the creation of the *Educational Multimedia Task Force*, grouping these and other related programmes such as MEDIA, presented for the first time an integrated approach, aiming at increased coherence and effectiveness of community actions.

The experience of this Task Force was very positive for the results of some of the over forty projects selected (one of them being, for example, EUN, the European Schoolnet) but somewhat disappointing for the integrated approach. Practical aspects of budgetary and administrative procedures took a larger than expected amount of the pooled resources. The Lisbon strategy is also here a watershed. The 'open method of coordination' calls for overall use of an integrated approach, not only between Community instruments, but also with Member States.

The first 'Lisbon action', and one fully representative of the new approach, was the global action plan *eEurope 2002* (Council of the European Union & Commission of the European Communities, 2000), setting ten key areas for common action, and concrete and quantifiable targets for each of them. Three of the ten actions target, directly or indirectly, education and research. Their main objectives are the improvement of the European telecommunication research network (the project GEANT, linking Member States, scientific networks); the connection of all European schools to the Internet; and training of a sufficient number of teachers. All three have almost been achieved (SEC, 2001b). This seems to prove the effectiveness of this approach, and a new global action plan, *eEurope 2005* (COM, 2002) has been approved recently. The main thrust of eEurope2002 was achieving simple connectivity; for eEurope2005 it is achieving quality connectivity broadband.

The *eLearning initiative* (COM, 2000) followed shortly after (May 2000), complementing and extending eEurope from an educational point of view. E-Learning identified four action lines: infrastructure and equipment; training for all; quality contents and services, and fostering networking, debate and cooperation at European level. These four lines have been proved adequate for structuring action and mutual information. In 2001, they were further developed in the *eLearning Action Plan* (COM, 2001c), a framework for increased coherence and synergy of the diverse EU programmes and actions in this field, and for a better structured dialogue with Member States, endorsed by a *Council Resolution on e-Learning* (Council of the European Union, 2001).

Also a direct request from Lisbon, *the Report on the future concrete objectives of education systems* (Education Council, 2001), adopted by the Education Council in February 2001, is the long-term strategy, with the same horizon as the general conclusions: 2010. The Report asserts three essential objectives for European education systems (quality, accessibility and connectivity), and defines each of them in a number of related objectives. Accessibility to ICT for all levels and all participants in education is one of them. It is also one of the three chosen as first priority, in the sense of demanding short-term actions.

The new paradigm of *lifelong learning* was the subject of a wide debate all through Europe, which culminated in the Communication *Making a European Area of Lifelong Learning a Reality* (COM, 2001b). This document sets the general principles for achieving this objective, such as, for example, valuing learning. ICT tools are seen as a positive contribution, in particular for adult education.

Last but not least, the *Luxembourg process* (European Communities, z.j.; COM, 2001a), for employment policy, is at the source of a number of documents where ICT skills are presented as an important factor for employability. The *Strategy for Jobs in the Information Society* (SEC, 2001a), also a 'Lisbon document', contains a well documented analysis of this aspect, and is the base for benchmarking one of the eEurope e-inclusion (SEC, 2001c; European

Communities, 2001) actions a sufficient number of public Internet access points, providing information and guidance on basic ICT skills.

A relevant consideration for the subject of this paper is that continuous professional development and training, including mastering ICT-enhanced tools, is essential for the higher education level. Universities thus face a new need, which may develop in future years to become totally different models of the now established ones. This could mean, for example, a lifelong relationship with the alma mater, which would become a source of guidance and of periodical 'brushing-up'. It could mean also different kinds of demand, not necessarily aiming at, for example, a degree.

2.3 The Role of ICT in EU Education Policy

The emphasis on current EU policy goes beyond the somewhat mechanistic view of provision of good infrastructure and equipment. Experience shows that the best equipment will stay unused if the right conditions do not exist. Sound pedagogical approaches, training and motivation of teachers and learners, basic organisation aspects such as adequate time and space are the key factors.

Higher education is no exception in this. 'Institutional' strategy for use of ICT for learning in universities is often absent, and efficient use depends also often on personal interest and on use of students' and professors' own time.

Equipment and connectivity are of course essential prerequisites. But the debate is now clearly centred on a good understanding of the real impact of ICT for learning and for teaching and for change. ICT tools are no more, and no less, than the most visible aspect of deep social and cultural change. The main objective of EU policy in this field is to exploit the potential of integration of ICT:
 – as a lever for change and adaptation;
 – as a catalyst for innovation;
 – as an enabler for lifelong learning;
 – as a support to employability;
 – as a bridge over the digital gap;
 – as a platform for European cooperation;
 – to foster a wide debate on the European knowledge society.

2.4 Full Integration of ICT in Higher Education: a Matter of Urgency

Integration of ICT is seen as having clear educational benefits, at large, and it is therefore an important objective for education systems. For Higher Education it is more than that, it is an urgent need first, because of the leading *role of Higher Education in education systems*. Technological and pedagogical research; training of teachers and trainers and development of new methods for including effective pedagogical use of the new tools; monitoring and assessment of school curricula; pedagogical and sociological studies; building of scenarios for change; these are all crucial roles of Higher Education. Delays or errors in their fulfilment will have a negative impact in lower education levels, and in social and economic development.
Equally critical and important is the role of Higher Education in society. The traditional functions of universities, creation and transmission of knowledge, and professionalisation, also become critical in a knowledge-based society.

Higher Education must face new professional needs, and must identify the new skills and competencies for building the knowledge society. In Career Space, a 'flagship project' carried out in cooperation between several General Directorates of the European Commission (Social Policy and Employment; Industry; Education and Culture) an interesting analysis on curricula for the 21st century has been developed. A striking conclusion is that soft skills and multicultural skills become as important as a solid scientific and technical grounding for new professionals.

The *role of Higher Education* for bridging the digital divide, the new social barrier of the knowledge economy, should not be forgotten. Building on understanding and research of what are the new competencies and skills needed for more and better jobs in a new economy, for critical and confident access to new forms of culture, for, what is most important, citizenship of the new society.

2.5 European Higher Education in the Global Society

European Higher Education has a long tradition as a hallmark of excellence. For centuries, European universities have been at the forefront of research, progress and culture. Europe is today at an inspiring moment, comparable to the Illustration, if not to the Renaissance. New ICT-mediated tools bring also

25

new opportunities for creativity, research and cultural and technological development.

There is a need to rise to the challenge, and to exploit new opportunities for enhancing quality and accessibility of European Higher Education. Not all the prospects are equally bright. Higher Education is affected, much more than other education levels, by globalisation phenomena. Competition becomes fierce, with new aggressive entrants. There appear unprecedented references to 'education as a commodity'. Our General Directorate is leading a reflection on the prospects for Higher Education in the oncoming GATS negotiations.

The answer should be a positive, assertive, one. The 'frontier-free' Europe should be at its most visible in the education field. Erasmus is one of the best European 'success stories', and a new 'Erasmus World' programme has just been adopted by the Commission, with the unashamed objective of attracting world talent.

2.6 EU Programmes and actions for ICT in Higher Education

All the above sets the context for community actions for Higher Education. It might be useful to recall that these actions, as belongs to a 'subsidiarity' sector, are led in close contact with Member States. There are periodical meetings with relevant participants, such as Directors for Higher Education or University Rectors; and there is also an ongoing discussion at the Education Committee of the Council.

It is interesting also to note that the main European initiative in the field of Higher Education, the 'Bologna process' initiated in 1999, fits in very well with the new 'post-Lisbon' approach, based on flexible and voluntary cooperation agreements, with a capacity to integrate intergovernmental and community initiatives.

European actions in the field of ICT for Higher Education can be structured, for the sake of presentation, in three different groups:

1 *A priority for existing education programmes*
   – Erasmus is the Socrates Action for Higher Education. Its main and best known action is the mobility scheme that provides for exchanges of students and professors all over Europe. In a few weeks, we will be celebrating the one-millionth Erasmus student. The extension of Erasmus has just been proposed, via the new Erasmus World programme.

–   Minerva is the Socrates Action for Open and Distance Learning and for new technologies. Around 75% of Minerva proposals come from the Higher Education sector, often articulating new partnerships with schools, cultural institutions, and other learning institutions.

2   *A priority for ongoing education initiatives*
The main one is the Bologna process, following the Bologna Declaration, signed by 29 Ministers for Education in 1999, and the Praga Communique, of 2001. As from last year, the Commission is a member of the Bologna monitoring group, and it is supporting studies, conferences and other information and dissemination measures about this process. Bologna´s main actions refer to definition of a common structure for university degrees (3-2 cycles); to mutual recognition (via ECTS, the European Credit Transfer System); to mobility; and to quality assurance. A key conference is scheduled for 2003, in Berlin.

3   *Other specific measures and actions*
In the first place, there are two new entrants: eEurope and its educational complement eLearning. The main impact of these new actions lie in their new implementation philosophy, more dynamic, open and flexible than traditional programmes. In view of their good results so far, both are developing new specific programmes (MODINIS and e-Learning), which will reinforce their coordination, monitoring, and leading roles.

The IST (Information Society DG, z.j.) part of the 6th Framework Programme will continue to have a specific action for education and training, which will focus on technological research. Embedded technologies, ubiquitous computing, and broadband applications are amongst its general priorities. The 2003 Work Programme should be adopted soon, and the first calls should take place in December 2002.

The research part of the 6th Framework Programme (Community Research and Development Information Service, z.j.) will also pay attention to ICT in education, and of its social and societal impact, under several of its priorities, in particular of the one on the new governance. There is increased cooperation between General Directorates for Research and for Education and Culture, which will coalesce in particular in the construction of the ERA (European Research Area).

27

2.7 The Specific Contribution of eLearning

The eLearning initiative and Action Plan were meant as pure coordination and 'synergy' action, building on existing programmes and instruments, with no own resources. However, the European Parliament has voted a specific budget under the 'innovative and exploratory actions' budgetary priority, to explore and foster development of e-learning in Europe.

The first two years of this approach have provided possibilities for several calls for proposals and for tenders, described below, and for the launching of a tender for building the eLearningEurope portal. Following the success of the calls (i.e. the large number of proposals received, and their different approaches, which seem to prove the need for a specific action) the European Parliament has asked for the development of a specific eLearning Programme, which should become not only a coordinator but a player in the eLearning Action Plan.

– *eLearning as an example of the open coordination method*
   There are two working groups with Member States: eLearning, and ICT-e, for the identification of common concerns; the definition of indicators and benchmarking procedures; and the identification of subjects or actions adequate for peer review

   There have been also a number of focus meetings, one of them with representatives of traditional and of open universities, to discuss and draw conclusions and recommendations on possible actions in the field of Higher Education. Participants recommended to further explore existing experience, and to better monitor and build on what already exists. Particular attention should be paid to competition issues, both at the global and European level.

– *eLearning calls for proposals: ICT in higher education projects*
   There have been two calls for proposals. The first one, in two phases, has focussed on the main eLearning priorities such as virtual universities, new learning environments or teacher training.
   Projects have been selected according to the following criteria: strategic approach; relatively large size (project financing was around 500000 Euro); demonstration potential; good perspectives for sustainability and good range of partnerships.
   Seven Higher Education projects have been selected. Six of them address different models for cooperation (LIVIUS, Cevu, MENU, GENIUS, DELOS, ICETEL, MusicWeb) and the seventh one, DELOS, is a blueprint for a European Observatory of eLearning.
   The second call, closing on September 30th, 2002, addresses two more

important issues: quality and media literacy. Some proposals related to Higher Education have been announced. External evaluation will take place during October; it is expected to finalise project selection and proceed to contracting within 2003.

– *eLearning strategic studies*

There have been two tenders for strategic studies, addressing areas where there is a lack of relevant decision-making information. On the first tender, closed in December 2001, and from which the results should become available in 2003, there was one specific study for higher education: Virtual models for European universities

Two others are also relevant for higher education: New learning environments, and Cultural institutions as new learning environments.

– *Fostering public-private partnerships*

The impact of eEconomy in industry (COM, 2001d) is very important; and the eLearning initiative has attracted a keen interest from industry. In May 2001, an important conference: the *eLearning Summit*, was organised by a group of top companies, mainly of the IT sector, to discuss in depth public-private partnerships. The Summit produced ten recommendations for development of e-learning in Europe (Z.a., 2001).

Following the Summit, a voluntary working group, *eLIG* (eLearning Industry Group) has been created by industry, to follow-up the conclusions, to provide a platform for dialogue and debate, and for the implementation of four pilot projects on connectivity; teacher training and leadership building; exchange of e-learning contents; and standards.

– *An European eLearning Portal*

With the ambition of becoming a single, user-friendly, entry point, and a European reference for the many 'e-learning flavours' The new portal is not designed as a competitor to existing sites, but rather as a reinforcement for all existing initiatives. It is also planned as a cooperation platform, facilitating contacts and partner-search between interested users. It should, finally, promote and highlight good practice in e-learning for education and training, and help the dissemination and exchange of ICT-based contents, tools and services.

29

2.8 eLearning: a New Programme?

The decision to present a new programme has been taken following the requests of the European Parliament. As mentioned before, it has been accompanied by a wide consultation of experts in the use of ICT for education and training.

The design criteria for the programme are as follows:
- focus on the most urgent issues;
- focus on tangible results;
- try to streamline and to simplify administrative procedures;
- work closely with Member States and key education stakeholders;
- foster pan-European dialogue and debate.

2.9 The Higher Education Component in the New e-Learning Programme

Higher Education is a first and foremost priority of the programme. In fact, it was at one point its only key action, as the reduced budget allocated to the programme called for focussing in a reduced number of significant actions. However, the choice of key subjects is by itself a declaration of political priority. It has been decided therefore to fix three priorities: higher education, schools, and fighting the digital divide.

For Higher Education, the idea is to provide an 'e-learning dimension' to the Bologna process. This would mean developing 'e-learning aspects' within its institutional cooperation and mutual recognition agreements, such as, for example, the inclusion of e-learning courses in the ECTS (European Credit Transfer System).

It would mean also the development of methods and tools for quality assurance; for developing schemes for virtual mobility, building on existing Erasmus exchanges; and fostering cooperation and exchange of good practice, and active dissemination and communication for a better accessibility to results.

2.10 Conclusions

Higher Education is crucial for a good and effective integration of ICT methods, resources and tools in education and training systems. The EU policy, following the Lisbon conclusions, calls on the social and academic responsibility of the sector to lead the way in this process of change, to build a European model for the knowledge-based society and economy, based in innovative and creative approaches, with a spirit of social cohesion.

References

COM (2000). *eLearning – Designing tomorrow's education.* nr. 318 final. 24.5.2000. The Commission's official webpage for the eLearning Initiative, (http://europa.eu.int/comm/education/elearning/doc_en.html).

COM (2001a). *Amended proposal for a* COUNCIL DECISION *on Guidelines for Member States' employment policies for the year 2002.* nr. 669 final. (http://europa.eu.int/comm/employment_social/empl&esf/news/emplpack2001_e n.htm#Guidelines).

COM (2001b). *Making a European Area of Lifelong Learning a Reality.* nr. 678. final. 21.11.2001. (http://europa.eu.int/comm/education/life/index.html).

COM (2001c). *The eLearning Action Plan – Designing tomorrow's education.* nr. 172 final. 28.3.2001.

COM (2001d). *The impact of the e-Economy on European enterprises: economic analysis and policy implications.* nr. 711. 29.11.2001.

COM (2002). *eEurope 2002: An Information Society for all.* nr. 263 final. (http://www.euro-pa.eu.int/information_society/eeurope/news_library/documents/eeurope2005/ee urope2005_en.pdf).

Community Research and Development Information Service (z.j.). *6th Framework Programme.* (http://www.cordis.lu/fp6/).

Council of the European Union, Commission of the European Communities (2000). *eEurope 2002: An Information Society for all. Action Plan.* (http://www.europa.eu.int/information_society/eeurope/action_plan/pdf/action-plan_en.pdf).

Council of the European Union (2001). *Council Resolution of 13 July 2001 on e-learning.* nr. 2001/C 204/02.

Education Council (2001). *The concrete future objectives of education and training systems.* nr. 5680/01 EDUC 18. 14.02.2001. Report from the Education Council to the European Council.

European Communities (2001). *Council Resolution of 8.10.2001 on 'e-Inclusion- Exploiting the opportunities of the Information Society for Social Inclusion'.* Official Journal of the European Communities. nr. C292.(http://europa.eu.int/comm/employment_social/knowledge_society/res_ein cl_en.pdf).

European Communities (z.j.). *The European Employment Strategy is based on four pillars: employability, entrepreneurship, adaptability and equal opportunities.* (http://europa.eu.int/comm/employment_social/empl&esf/ees_en.htm).

Information Society DG (z.j.). *The Information Society Technologies (ist) Programme.* Part of the 5th Framework Programme for Community research. (http://www.cor-dis.lu/ist).

SEC (2001a). *Benchmarking Report following-up the 'Strategies for jobs in the Information Society'*. nr 222. 7.2.2001.

(http://www.europa.eu.int/comm/employment_social/knowledge_society/bench_en.pdf).

SEC (2001b). *eEurope 2002 Benchmarking, European youth into the digital age*. nr. 1583. Commission Staff Working Paper.

SEC (2001c). *e-Inclusion – The Information Society's potential for social inclusion in Europe*. nr. 1428. 18.9.2001. Commission Staff Working Paper.

(http://www.europa.eu.int/comm/employment_social/knowledge_society/eincl_en.pdf)

Z.a. (2001). *European eLearning Summit*. La Hulpe, 10-11 May 2001, (http://www.ibm-weblectureservices.com/eu/elearningsummit).

3 EU Policies for e-Learning as Seen from the World of Higher Education

Peter Floor, Coimbra Group

3.1 Introduction

3.1.1 *Coimbra Group*

The Coimbra Group of Universities is a network of, at present, 35 universities aiming at stimulation, initiation and support of multilateral international collaboration among traditional European Universities in areas of academic and cultural activity where added value may be expected (see www.coimbra-group.be). The Group started in 1985 at the initiative of Mr Simon-Pierre Nothomb of the Catholic University of Louvain and has proved its value from the early days of the ERASMUS Programme onwards. In the early nineties the Group initiated a discussion among its members about the potential of the new Information and Communication Technologies (ICTs) in Higher Education and their importance for traditional, research-oriented universities. This process resulted in a series of EC-supported projects that gave staff and students from interested universities across Europe an opportunity to get hands-on experience in preparing and testing common virtual modules as innovative parts within their local classroom taught courses (mixed mode; see www.dipoli.hut.fi/humanities/) and yielded interesting building bricks for further developments (see www.dipoli.hut.fi/humanities/hum-pub.html).

3.1.2 *European ODL Liaison Committee*

The increased focus on ICTs, both in regional, national and European policy-making and in education itself has prompted institutions and organisations working in the field to form networks to benefit fully from the potential of a European dimension in their activities and to respond to increasing demand from their network members in this area. Representatives from a number of these networks, among which is the Coimbra Group, agreed that a common platform would be an added value for each of the networks and they decided in June 1998 to establish the European ODL Liaison Committee (LC; see www.odl-liaison.org). The Committee meets regularly to discuss matters of common interest among its members and with representatives from the European Commission (e.g. to discuss the creation of the MINERVA action within the SOCRATES Programme, and the eLearning Action, see below).

3.1.3 This contribution

After the launch of eEurope by the June 2000 Lisbon European summit, eLearning became one of the major areas of action to implement eEurope, education being the principal factor for the professional and cultural training of the men and women who would have to put Europe in the frontline position in a globalised, digitalised world. The Liaison Committee and its constituent networks reflected intensively on the kind of tools needed by European, national and regional authorities to give the most effective incentives to the world of education, and on many aspects of implementing eLearning, for instance, the capabilities of the different sectors in the education system to perform the attributed roles successfully. This contribution summarises the results of two activities, one initiated and coordinated by the Coimbra Group, the other by the European ODL Liaison Committee, to explore the perspectives for fruitful and sustainable innovation that are essential for achieving the goals of eLearning. Both were mainly realised during 2001.These results are placed in the less optimistic social-economic context of mid 2002 and the question is raised whether the changed conditions call for revision of earlier established policies. In the concluding section some arguments against a possible need for revision are given. This leads to a strong appeal to all involved parties to adopt new ways of cooperation aiming at concerted action for eLearning implementation.

3.2 The Socio-economic Context[1]

Roughly speaking, in the time leading up to the June 2000 Lisbon European summit, the world lived in a kind of euphoria. Unprecedented economic growth was said to be a more or less permanent feature of the post-Cold War/post-industrial era and was believed to be able to solve the remaining problems in the world.

ICTs were thought to be a major driving force of the New Economies. Much political attention was focused on how to seize a proper share of the wealth created, in our case for the European Union in competition with the United States and Japan; and also on how to ensure a proper share of political, economic and cultural influence in the rest of the world. The industry involved played an active role in all this.

The world of education witnessed these developments with mixed feelings, all of its different sectors in their own way, and within these sectors with some strong regional differences too. On the one hand there was an institutional and personal recognition, and resulting drive, to use the ICT potential to serve their goals and if possible to gain relative advantage over others. On the other hand many stakeholders felt themselves too much under all kinds of government pressures to address rationalisation, quality issues, access for underprivileged groups of the population, improved teacher training, follow-up of the Bologna Declaration, etc. and could not cope with (seemingly) unrelated aspects of ICT implementation on top of that. Individual teachers feared that ICTs would present themselves as a 'camel's nose' and result in a permanent threat (financially, organisationally and pedagogically) to current activities, or even academic positions.

35

For the latter category the dot.com crisis may have come with a sigh of relief and provoked exclamations that they had always known that the ICT revolution would not materialise, that their reluctance was justified. Moreover, the present economic crisis may now result in restrictive policies that will affect all sectors of education. So, there may be movements to forget about ICT for some time and solve immediate problems first. In those institutions (like universities) that have some degree of self-government, this may also affect the limits of freedom (in terms of acceptance by peers) of the leaders to introduce ICT matters in institutional strategic processes.

Does this general and maybe in some respects exaggerated view mean that we had better set aside earlier, or even the most recent, policies and actions? In a psychological sense maybe yes, but in all other respects certainly not. Maybe these developments could even result in better opportunities to bring EU/national policies and the world of education together. These issues will be elaborated in the Conclusions after having analysed aspects of the present situation in the following two sections.

3.3 Higher Education Strategic Response Capability: the HECTIC Project

Political objectives for education/training and eLearning have been proposed by the European Commission and were accepted by successive Councils of Ministers. Both the Commission and national/regional authorities now have an obligation to try to translate these objectives, which are of necessity of a fairly general nature, into practical measures which result in implementation of the objectives, and within the time limits set. eLearning did not come out of the blue and in earlier years the generation of a synergy with sustainable effects had not resulted from its forerunners (ODL, etc) despite serious efforts on both sides (EC/governments and education institutions) and considerable input of EC project grants and national/regional programme funds. As appears from other chapters in this book, this of course varied from country to country – some (e.g. United Kingdom and Finland) made more progress than others but overall less than was hoped – and from institution to institution (e.g. Twente, Leuven, Lund and Edinburgh).

The Coimbra Group, dedicated to promoting synergy between the Group members, among institutions of higher education, and between these and European authorities, felt that it might be able to contribute to a realistic approach to the eLearning implementation by analysing the general lack of sustainable progress in the past years. Together with some other networks and institutional partners it felt well-positioned to consult the collective experience. It stated in its project proposal: 'We therefore feel that, better late than never, a serious attempt should be made to gather the views from leaders in a number of major, often strongly research-based, conventional universities on how such institutions could be motivated and enabled to make the leap into being digital and what policies at governmental, and especially European level would be instrumental to create the right conditions for these changes. The proposed project aims at mobilising these views, at a thorough solution-oriented discussion of them, searching for valid cures of the problems encountered, testing these on a wider audience for validity and then presenting them as recommendations to the European Commission'. The Project Higher Education Consultation on Technologies of Information and Communication (HECTIC) was approved by the Commission as a one-year project under SOCRATES/Complementary Measures, to be done roughly in the year 2001.

The final HECTIC report: 'European Union Policies and Strategic Change for eLearning in Universities' is available via the Coimbra Group web site[2]. The

Executive Summary summarizes its main considerations as follows:

–   The revolution in information and communication technologies is having significant effects on higher education and there are many examples of its impact on teaching in universities and colleges. The European Commission is well aware of this significance and of the potential of these developments and has created an eLearning Action Plan with a 10-year horizon to implement policy objectives approved by European Council meetings. At the same time the lack of sustainable innovations and of diffusion of eLearning experiences outside the circles of directly involved academics is a concern of many observers, both in the European Commission and in Higher Education.

–   While the new ICTs are having a variety of direct effects on teaching and learning in universities, there are also a number of other important factors having major influences on higher education which have to be taken into account when considering how the new technologies can be most effectively utilised. The processes in implementing the Bologna Declaration are having impacts on the development of curriculum structures and quality control attitudes and procedures. The rise in lifelong learning and the widening of access bring in new learners with different previous educational experiences. At the same time, the increase in the total demand for higher education and the consequent creation of a bigger market is attracting new providers, many of which are 'for profit' organisations. This large and diverse market is expected to produce a more heterogeneous range of higher education institutions as universities seek to identify strategically specific niches in which they hope they can maximise their chance of being successful.

–   The extent and range of new expectations on universities also create important challenges for the institutions themselves in implementing eLearning. Among the most important factors is the question of staff motivation. What recognition and rewards can be used to persuade academic staff to change their teaching methods and incorporate new technologies? What structural and legal changes might facilitate such changes? Associated with these issues is that of the effectiveness of leadership within institutions, not only to stimulate, enable and reward the uptake of eLearning but also to create a level of strategic thinking and planning for the university as it adapts to both external pressures and internal opportunities.

–   The Educational Policy of the EU has three main objectives, which concern:
    –   increasing quality and effectiveness of education and training
    –   facilitating access of all to education and training
    –   opening education and training to the wider world

HECTIC's view is that eLearning is relevant to all three objectives, but also noted that significant perceptive, conceptual and practical frictions exist between the two worlds of policy objectives and educational developments. These need to be bridged to generate real synergy and thus relevant implementation of eLearning serving also institutional strategic priority achievement. Furthermore, virtual mobility for students and staff and for the concept of a European virtual university should be stimulated. ICTs clearly have great potential for virtual mobility for students, but it should complement and not replace physical mobility. The development of virtual university activity on a European scale could best be achieved by building on existing activities.

- Achievement of the full potential of eLearning requires action by a number of players. A number of specific recommendations were made, some of which are addressed to institutions and their associations and some to the European Commission. Among the former are those relating to strategic leadership and management; while those to the Commission include proposals for the improvement of ICT infrastructure across Europe and the incorporation of eLearning in the development of the Bologna Process. But the majority of the recommendations require coordinated action by the Commission, the higher education system and, in some cases, the relevant national governments or their agencies. Among the topics contained in these recommendations are:
  - work with the European Commission to develop coordinated virtual university activity across Europe;
  - coordination across Europe of European and national objectives in eLearning;
  - review of funding mechanisms and procedures;
  - investigation of multiple sources of funding for eLearning;
  - the need to evaluate current experience and disseminate good practice;
  - promote the spread of virtual mobility;
  - research into the effectiveness of eLearning;
  - consideration of the impact of Intellectual Property Rights.
- A major recommendation was made for an overarching framework to drive forward the recommendations which need coordinated action: the creation of a standing body with representation involving the Commission, European university associations and students. Practical ways to engage national governments and agencies on relevant topics would need to be found. Not only would this body have the responsibility of considering how the specific recommendations are to be implemented, but it would also be able to monitor all further developments in eLearning for higher education

and make appropriate additional recommendations for concrete actions.

– The HECTIC groups agreed that the discussions, involving both the Commission and representatives of the higher education world, had been extremely valuable. They wish to see the above-mentioned mechanism to ensure the maximum benefits for the promotion of sensibly selected eLearning development across Europe through a close interaction between the Commission's implementation of EU policy on the one hand and the dynamic developments occurring in higher education institutions on the other.

Furthermore, the HECTIC report elaborates a wide range of aspects related to the topics listed above. We shall highlight some which are specially relevant for the purpose of this contribution.

3.3.1 Higher education challenges

The HECTIC report states that, in addition to external pressures arising from wider developments that universities would have to address anyway, the fact that political leaders embrace the potential of ICTs to open new markets, to create new opportunities for education and training, and to assist in reducing geographical and social exclusion, creates a pressure in itself (HECTIC report, chapter 3). A number of challenges which university leaders have to address were identified. These can be grouped together under the headings of cultural/managerial, technological, and educational. A further distinction is made between challenges of a more strategic nature and those affecting implementation of (almost any) strategic decision in this context. University leaders, unavoidably confronted with these challenges, will have to consider which would be the most suitable response for their institution. Given the complexity and interrelation of the many factors they have to take into account together with the consequences of eventual adaptations in their universities, this consideration will have to lead to no less than a set of strategic choices positioning the university in its proper market niche. HECTIC participants estimated that only a minority of the European institutions have completed this process. Moreover, in many cases, rectors and their senior colleagues do not feel well-equipped to face major issues coming up in higher education together with their daily concerns and the internal structure of responsibilities (HECTIC report, chapter 4).

3.3.2 *Need for institutional strategic response*

The following elements of a strategic process can be distinguished:
- strategic discussion , involving key persons at all levels of the institution, leading to setting of priorities which define the institution's preferred niche in the education arena;
- taking care of an adequate technological infrastructure, also reaching out to the students' living rooms, including reservation of funds for renewal;
- creation of an enabling environment and staff incentives involving:
  - coordination of existing expertise on pedagogy and technology for new learning plus media and library facilities in a coordinated resources and support structure accessible for the whole university community and filling of gaps in expertise if they appear;
  - ample opportunities for staff development and staff training at all levels;
  - staff rewards and normal career perspectives for those staff who (temporarily) engage themselves largely in eLearning implementation at the risk of, for example, reduced research output;
  - reservation of some seed money, also for temporary employment of young staff to assist in the curricular reform activities.

HECTIC adds that the actual implementation of strategic choices will usually take more time than the period of office of Rectors and other university leaders so care should be taken to guarantee continuity. It lists a variety of issues and factors to be taken into account when discussing strategy.

3.3.3 *European policy issues and higher education ability to respond*

The main issues at European policy level concern:
- the Education Council's Concrete Objectives of Education and Training Systems;
- the Bologna process towards the creation of a European Area of Higher Education;
- the European Commission eLearning Action Plan;
- the European Commission Memorandum on Lifelong Learning.

On higher education's ability to respond to EU objectives, HECTIC stated that: 'When stepping down from global observation to daily practice it becomes clear that a great number of minor inconsistencies in the chain of interdependencies make it very difficult for university leaders to respond fully, and in the speed required by external authorities, to the challenges they have to face' (HECTIC report, chapter 6).

In brief, these inconsistencies are about:
- creation of enabling institutional environments to motivate staff for institution-wide responses;
- short-term national/regional political interests that may overshadow what would be best to assist universities to achieve goals also wanted by these authorities;
- the lack of fine-tuning in making available financial means together with setting of new government priorities;
- disregard by authorities of time needed by universities to react effectively;
- legal or higher-level formal constraints which may counteract required policy implementations;
- lack of sustainable results of EC programme funding (see next section of this contribution);
- governments' lack of initiative to make international networks instrumental for their goals;
- greater awareness of issues at stake by ministers or ministries of economic affairs than by ministries of education;
- estimated future increase of regional and EU roles in HE policy making, specially for ICTs;
- potential initiating role for EC in finding practical and fair ways of engaging private money in the reform of university education.

After this specific discussion the report concludes that: 'The points of view sketched above all give evidence of frictions in communication, in understanding each other's positions and span of control, in not well-understood mutual expectations, etc. If this situation is not recognised by higher authorities and taken into account when developing policy matters, major policy operations like the implementation of ICT in education risk falling far short of their objectives. This cannot be remedied by simple changes of procedures and attitudes of the participants, though both are very important. Major policy processes are too complicated to be approached in a monoschematic way. A flexible approach with constant and committed participation of all involved parties will be needed'.

3.3.4 *Permanent and committed mechanism for dialogue recommended*

Hence the HECITC recommendations focus on the need for the creation of a mechanism for dialogue between the Commission (and through it national/regional authorities) and the higher education sector. In the HECTIC view, this necessarily permanent and committed interaction is expected to result in the reduction or removal of the most serious obstacles for university leaders that prevent them responding effectively to political goals which they will usually share in principle but feel handicapped to engage with at present.

3.4 Effectiveness of EC Tools to Achieve Major Changes: the European ODL Liaison Committee Policy Paper

HECTIC focused on higher education, and furthermore was deliberately designed in such a way that the Programme structure of the European Commission and the processes around European projects would not draw attention away from higher order, more political considerations. In the European ODL Liaison Committee, besides higher education, many other sectors are working together. All these networks are under constant pressure from their members to give high priority to critical but honest and constructive discussion of day-to-day project practices, so the Liaison Committee decided that it would take up this issue and come up with practical recommendations. The resulting policy paper 'A Framework for European Commission Programme Funding / Analysis, Recommendations and Proposal for Discussion' was released 4 February 2002[3], after having ascertained that the text could indeed be a useful starting point for discussion with representatives of the European Commission. The most salient points of the paper are summarised below.

It should be stressed that the LC discussed EC Programme funding in so far as it is designed to realise major political objectives like, at present, eLearning as part of eEurope, and in full width, involving all relevant DGs. Nevertheless, part of the observations may well apply to the whole of Programme practice and the LC thinks that the European Commission would lose an opportunity if it did not take these into account in its habitual monitoring and evaluation of Programmes.

3.4.1 *Analysis*

The LC policy paper analyses three – in reality interrelated – aspects, the conclusions of which, quoted from the paper, are the following:

– *Political Objectives of Programmes*

The Liaison Committee agrees with the European Commission that the European Union has to play a stimulating and dynamic role in sectors of activity where Europe-wide effects and interests are at stake. However, it comes, to the conclusion that it will be necessary to redesign fundamentally the mechanisms used to implement such general and overarching policy objectives if the Commission wishes to be effective and sustainable in this complex and huge task.

– *Programme Structure*

  1 The present Programme structure is not capable of ensuring the uptake of interesting bottom-up project results and otherwise obtained innovations in the mainstream of European educational practice. The structure itself should not be blamed for that.The cause is that insufficient attention has been given to the fact that implementation of major political objectives needs other, more top-down oriented process developments for which adequate stimuli do not exist now, either at the national or at the European level.

  2 Within the present Programme structure so many complicating factors have been introduced that careful evaluation and revision, after redefinition of the objectives, seems to be inescapable.

– *Handling of Programmes in Practice*

The revision of the Programme structure should lead to revision of the way of handling projects, from calls for proposals to final acceptance of the financial justifications.

3.4.2 *Recommendation*

The Liaison Committee recommended not only the subject matter to be taken into account but also the actions that need to be taken (in a coordinated way that is capable of generating powerful synergy: contacts with political persons/bodies, both European and national, which have to play a role in the adoption of a revised structure; publicity; discussion with the members of the networks; etc.). The subject matter would comprise:

– Agreement on the need for new instruments for research and implementation of major policies in education, including other, closer and

more interactive relations with Networks/organisations of stakeholders;
- Definition of the objectives and main characteristics of these new instruments;
- 'Cleaning' of the present Programme structure of elements that were mainly introduced to let the Programmes play a role in policy implement-ation; if this did not appear to be effective, reconsideration of the funded programme approach;
- Parallel to this cleansing: streamlining of the Programme rules, regulations and practices.

The result aimed at should be no less than to have in place the best mechanisms to implement any agreed policies. Since it would not be realistic to suppose that the Commission would be ready to engage in a general discussion of this nature, the LC proposed to focus on one specific area within the whole domain of education, i.e. eLearning, for which the LC has been created and for which major policy development is under way. It could serve as a test bed for ideas and may be followed in other areas when windows of opportunity appear.

44

3.4.3 Proposal focussed on eLearning

In its proposal, almost fully quoted below, the Liaison Committee takes the point of departure that the European focus should be on content rather than on funding mechanisms and project management. Moreover, essential changes are not only effected by EU funding programmes (for example, in higher education, the process to implement the Bologna Declaration). Basic issues should be identified and funding should be linked to core business. An attempt should be made to use the principle of subsidiarity as an asset instead of a limitation at the European level. The proposal of the LC therefore consists of three parts:
- Conceptual strategic ideas
- Relation European Commission with the world of education
- Instruments, including mechanisms and sources for funding.

3.4.4 Conceptual strategic ideas

In the analysis section of the policy paper it has been argued that the present organisation of programmes leading to sequences of pilots is inadequate to achieve major educational policy goals, since this practice does not lead to sustainable results which, from within and bottom-up, effect institutional innovation and change. This lack of connectivity was said to originate from the lack of stimulating or enabling conditions in institutions operating in the field of education. The character of these conditions may vary according to the sector and kind of education and should be better understood. For instance, in the case of Higher Education, the HECTIC project has made it clear that institutional leaders (rectors, etc.) are insufficiently equipped to carry through fundamental processes of strategic definition of the market niches that fit their institutions best and implement the changes leading to taking those market positions. The LC claims that the networks represented in the LC are competent to analyse this aspect in the different sectors or kinds of education and the networks are ready to commit themselves to doing this within the time limits set in further discussions with the Commission. This definition process will lead to sectoral goals, like in the case of Higher Education, the creation of enabling environments in individual institutions. With these goals in the picture, long-term action lines with systems of related projects to achieve these goals can be designed in committed but open cooperation.

3.4.5 Relation European Commission with the world of education

45

Input will be needed, and given if parties agree to cooperate, from outside the European Commission to define the goals for achieving sustainable results in institutions for education and training. Along the same line of reasoning it will be clear that a similar input will also be essential in the next phases: programme design (definition and setting of priorities), and implementation (mechanisms, rules, monitoring of progress). Umbrella organisations can provide this input. They have the overview, can inform or consult and get feed-back from their members, can make manpower available in such a way that the interests of the organisations and their members as proposers of projects in ongoing programmes are not harmed (agreement on code of practice/conduct needed), and they can monitor progress, select what are, in their opinions, the best things done and care for continuity. They can assist in the creation of mutual synergy between innovation driven activities and research. Finally, they can also draw on other communication lines of their members with national authorities than can the Commission. When compared with the prevailing actual situation this means an essential position between the Commission and its relevant DGs and offices and the target groups, and a content-fed enrichment of the tasks now done by the TAOs. The networks in the LC are umbrella organisations and declare themselves willing to engage in agreed operational activities.

3.4.6 Instruments, including mechanisms and sources for funding

When discussing possible instruments, the LC concluded that improvement is not 'simply' a matter of larger projects. The Commission should be able to rely on a larger diversity of means to encourage and enable creation of ideas and incentive work. The approach should be changed; there is a need for diversification of tools supporting development, and, not less, of the way to use them.

In other parts of the policy paper, instruments have already been mentioned or implied: adoption of the principle of cooperation between the European Commission and the networks to implement objectives and goals in a committed way; investigation of the reasons for the lack of connectivity that hinders institutional take-up of bottom-up project outcomes; definition of sectoral goals that will result in relevant implementation of European Policy Objectives; long-term action lines consisting of skilfully combined project activities; objective-driven proposals (not money-driven proposals); input by the networks and participation of persons from the target sectors of education in relevant parts of the management of agreed EC actions; a process design and time schedule that satisfies the needs of the EC (and national authorities) to show robust progress at the same time coordinated with other ongoing national and European political actions and reforms, so that synergy results, down to the level of institutional leaders. In short: a process that challenges rather than commands.

In order to realise all this we shall also have to look for appropriate financial instruments that can be applied in selected cases to show that it will pay off for individual institutions (either individually or in consortia) to do more than the average. This could also mean that systems of multiple funding are developed in this domain and that funding of agreed activities may come from a combination of them or from a single source. In this motivating way the EC principle of co-financing may be bypassed in appropriate cases. We do not exclude a role for the Commission, or nearby the Commission, of a venture capital broker (private sources or European Investment Bank, etc.) to bring more flexibility into the process of educational innovation and reform.

3.5 Conclusions

3.5.1 *No reasons for revision of European policies*

In the last paragraph of the Social-Economic Context section above, the question was raised as to whether the economic decline and the dot.com crisis would be arguments to rethink fundamentally European policy objectives established in better times. Tentatively the position was taken that the new conditions might actually favour the bringing together of EU/national policies and the world of education. I shall now try to argue this position. The arguments are of various nature and in random order.

- The EU policies call for reforms in education because Europe, as the government leaders see it, needs a large workforce of differently trained and better equipped, flexible professionals. Even if these structural reforms had been realised now, it would take years before alumni of the reformed system would become available. Without these reforms already done, full results can only be expected in about 10 years. During the economic boom, increasingly serious deficits on the labour market were forecast, so the slow-down of the economy may provide better chances of a balanced development of supply vs. demand of skilled workers.
- During the boom, almost any graduate of any study easily found employment. In the new situation this will no longer be the case. Universities will have to care more about how to give best career perspectives to their students. This may lead to better prospects for acceptation of institutional strategic processes.
- European Commission and Governments will feel greater pressure to take measures in support of a rapid reform of the education systems as part of their contributions to overcoming the economic dip.
- European Commission, Governments and European industry may feel greater need to generate synergy in order to occupy strong positions on the changing global market.
- The Bologna process is having enormous consequences for higher education systems in most European countries. From many sides it has been stressed that not fully including eLearning in the Bologna process would be a serious omission.
- The world is not only facing economic competition. Europe can exploit certain aspects of the present crisis to its advantage as it can insist that it has certain unique features which can provide alternative concepts, as for instance, but certainly not only, in business administration, that may suit

other parts of the world as well. It may require reconsideration of existing concepts and alliances. Education and research are to give a solid background to this European social-cultural consciousness. Education has great opportunities in the dissemination of courses that reflect this way of thinking. European authorities will be expected to, and can, catalyse this process.

3.5.2 eLearning policies urgent, but other approach needed

In my opinion there is a stronger need to pursue the established policies for eEurope/eLearning rather than to reduce their urgency and impact, or review them, so our analyses and recommendations of ongoing processes as done in HECTIC and by the Liaison Committee are still relevant. Based on earlier decisions, the European Commission is now preparing incentive actions involving financial support, elaborating the eLearning Action Plan into an eLearning Programme. As a consequence the need (as established by the HECTIC project) to create favourable conditions for university leaders to carry through strategic re-orientations remains a most urgent issue which should be addressed by the university leaders collectively in dialogue with European and national/regional authorities. At the same time these major political issues demand EC support partly targeted at other aspects of university activity with appropriately adapted rules and procedures, as argued in the European ODL Liaison Committee paper. Once more: the best way to achieve sustainable results will be to design effective and innovative approaches in dialogue. Given the advance of the eLearning Programme development this is a most urgent need as well. Of course, the dialogues proposed in the HECTIC report and in the LC paper are not the same but there has been sufficient conceptual overlap when preparing both papers that smooth implementation could be ensured when the Commission is ready for it.

3.5.3 Epilogue

The proposed changes in attitudes and procedures will by no means be easy to achieve. The world of Higher Education has no reputation of uniting forces when necessary and keeping ranks closed as long as necessary, especially when getting nearer to the vested interests of its members. The European Commission has countless personal contacts and may get any advice it wishes to get, so why embrace just these rather complicating proposals? Moreover, synergy between involved Directorates General seems hard to bring into practice, despite important initiatives by many participants. As argued in this contribution, the challenge of the ICT policies now at stake is such that all parties are expected to feel the professional obligation to look for the best possible lines of action and set aside some habits that may be justifiable in other circumstances, but would be counterproductive here. ODL-related networks and university associations and many experienced senior persons in them are ready to respond to this challenge.

Acknowledgement

Dr. Jeff Haywood of the University of Edinburgh critically read the draft and made very useful suggestions. His kind assistance is gratefully acknowledged.

Notes

1   This section reflects the author's personal views.
2   www.coimbra-group.be. The electronic version may also be obtained from the Coimbra Group Brussels office via Ms Noelia Cantero: cantero@coimbra-group.be . Publication of a printed version is under consideration.
3   Full text can be obtained at www.odl-liaison.org.

4 Loosely Coupled Policy Links – ICT Policies and Institutional Dynamics in Norwegian Higher Education

Peter Maassen & Bjørn Stensaker, Norwegian Institute for Studies in Research and Higher Education, Norway

4.1 Introduction

Over the last decade ICT has become an important policy topic in higher education, at the supra-national, national, and institutional level (Green, 1998; Geloven *et al.*, 1999; Collis & van der Wende, 1999; Pedro, 2001). This statement, which is so obvious that it almost can be regarded as a cliché, covers in itself a complex set of relationships between these three policy levels. This includes the complexities that characterize all regular public policy processes, as well as the specific nature of the ICT-(r)evolution, the consequences of which we are only beginning to understand. An additional complexity with respect to higher education is that higher education institutions are both ICT innovators and ICT implementers. They have an important input in the ICT innovations, especially with respect to the technological dimension, while they are expected to be affected by it from the political side without losing their specific character nor their role in society. Castells has expressed this interesting tension the following way:

> *The development of electronic communication and information systems allows for an increasing disassociation between spatial proximity and the performance of everyday life's functions. However, schools and universities are paradoxically the institutions least affected by the virtual logic embedded in information technology, in spite of the foreseeable quasi-universal use of computers in the classrooms of advanced countries. But they will hardly vanish into the virtual space. In the case of universities, this is because the quality of education is still, and will be for a long time, associated with the intensity of face-to-face interaction. Thus the large-scale experiences of distance universities, regardless of their quality, seem to show that they are second-option forms of education which could play a significant role in a future, enhanced system of adult education, but which could hardly replace current higher education institutions. (Castells, 1996: p. 397; see also Dokk-Holm & Stensaker, 1999, for a similar discussion in a Norwegian context).*

In this paper we will discuss the interaction between governmental ICT policies and institutional responses in higher education on the basis of the experiences and practices in Norway. For that purpose we will examine the relevant Norwegian governmental policies with respect to ICT and higher education. We are, amongst other things, interested in the governmental interpretation of the concept of the knowledge society, the nature of the governmental expectations with respect to ICT and higher education, the policy instruments the government is using for realising its policy goals, and the role the government has reserved for institutional initiatives and actions in realising ICT policy goals. In addition, we will present and discuss the opinions of institutional leaders concerning the use of ICT in their institutions in general, and their attitudes with respect to the governmental ICT policies in particular. In doing so we will analyse the extent to which there is a match between the governmental and institutional positions concerning ICT and higher education. This analysis will help us to shed light on the issue of whether the further development of ICT in Norwegian higher education will be mainly a top-down process, a bottom-up effort, or a combination of both.

4.2 Data and Method

The paper draws on three sources of data. First, it discusses two White Papers on higher education in Norway, which include specific governmental policy approaches concerning the use and implementation of ICT in higher education. Second, it relies upon data from the international survey on ICT adaptation in higher education (see also chapters 8 and 9). These data are used to analyse the opinions of important decision-makers in higher education institutions concerning the ICT strategies, and their implementation at their own institutions. Twenty top institutional decision-makers from 17 of the 36 higher education institutions in Norway responded to the survey that took place in December 2001 and January 2002. Third, in spring 2002, we conducted five institutional case studies (four colleges and one university) in which we analysed the state of the art concerning the use of ICT in teaching and learning. In each institution, strategic policy documents and action plans for implementing ICT were analysed together with evaluation reports on the actual use and status of ICT. In addition, 10 to 12 interviews were carried out at each of the five institutions. Interviewees were mainly those responsible for the implementation of ICT, including ICT support staff at the institutional level, deans, and representatives from the central administration. Overall, it can be argued that the data provide for both a general overview and a detailed examination of the actual use of ICT in teaching and learning in Norwegian higher education.

4.3 Norwegian Higher Education – a Short Introduction

Before examining the links between policy initiatives and institutional actions, a short, general introduction is needed on Norwegian higher education. Higher education in Norway is mostly a public affair. Even though there exist private higher education providers, such institutions only concentrate on niches in the higher education market. The major exception to this is the private Norwegian School of Management BI that provides business studies through a number of smaller regional institutions, and MBA and PhD-studies at their main campus in Sandvika, Oslo. All four 'traditional' universities in Norway are public, with the largest, the University of Oslo established in 1811, having over 30,000 students. The higher education sector consists, in addition, of 26 state colleges spread throughout the country. Together, around 180,000 students are currently (2002) enrolled in higher education.

The Norwegian universities clearly 'match' what Clark (1983) has called the 'continental' model of higher education. The Humboldt ideal has been quite strongly articulated in the universities. Thus, central characteristics are departments with much power, weak institutional leadership, input-based public funding, and much emphasis on traditional lectures when it comes to teaching methods. The college sector offers shorter studies (usually up to three or four year) in engineering, health and social work, business education, but also in more traditional disciplinary fields, such as sociology, political science, and language studies. The college sector was reorganised in 1994, merging 98 former regional colleges into the 26 new state colleges. The amalgamation created many multi-campus institutions with quite extensive distances between the various campuses.

Due to the scarcely populated countryside and the often huge distances between the higher education institutions and potential students, distance education has for decades been a trademark for higher education. Norway was, for example, the first country to introduce legislation on quality control of private distance education providers in 1948 (Ljoså, 2002: p. 14). Distance education has, however, mostly been seen as a matter of necessity, and not as a 'proper' way of acquiring higher education (Rekkedal, 2002: p. 29). Nonetheless, distance education has traditionally been an important 'testing ground'

53

for new teaching and learning technologyincluding, in the last decade, ICT. At a few Norwegian universities and colleges one may witness an interest in this way of providing and delivering education, but the attitude has mostly been that the use of new technology was not something for the 'traditional' students on campus (cf. also the above quote by Castells).

During the last decade, various green and white policy papers emphasising 'service management attitudes', more 'focus on students individual needs' and 'flexible delivery solutions' throughout higher education, have contributed to a new political agenda for higher education (Rekkedal, 2002: p. 30). Not least, more emphasis on the need for higher education institutions to attract external funding, for example, through commissioned courses, and lifelong learning schemes play an important part in this development (Brandt, 2001).

4.4 ICT and Higher Education as a Policy Issue

The realisation that the information and communication technology (ICT) is drastically changing our world became visible in the social science literature at the end of the 1980s and beginning of the 1990s. Prominent scholars discussed especially the effects of the ICT development on national economies and social structures. They argued that, concerning the economy, ICT allows for the development of a truly global economy. In this global economy national borders, traditions, reputations, and policies lose their importance. Production is flexible and can take place anywhere in the world. Traditional mass production is gradually disappearing. The underlying structure of this global economy is a network kind of organisation, made possible by ICT.

Complementary with a global economic integration one can observe a growing social disintegration or fragmentation. Traditional social relationships, such as the relationship between employer and employee, and family life organised on the basis of a marriage, as well as social institutions, e.g. full-time employment, and unions, are being threatened. This social disintegration is, amongst other things, being referred to as the individualisation of society.
The scholars in question also became involved in national governments' attempts to understand the ICT developments and the translation of this understanding into national policies. Their position with respect to ICT at that time can be summarised as follows: 'The world economy is becoming more competitive, more global, and increasingly dominated by information and communications technology' (Carnoy et al, 1996: p. 1). This statement by

Carnoy and his colleagues was made in 1992 in the framework of advice to the Yeltsin government in Russia on political economic policy. As indicated, it reflects in a sense the 'time spirit' of 10 years ago. There was still a strong belief in the 'new' economy. In addition, the authors were convinced that ICT development is one of the most significant technological revolutions in human history. Finally, they discussed the possible consequences of the information technology revolution for the nation-state with the following conclusion: 'The role of the nation-state in creating an innovative society is absolutely crucial to the well-being of its citizens in the information age' (Carnoy, 1996: p. 91). While the belief in the newness of the new economy has withered since then, there is still the conviction among scholars as well as politicians that the ICT developments are of crucial importance for the economic well-being of nations. The knowledge economy has become an important policy topic, with ICT as its engine.

From this perspective it is obvious that higher education deserves appropriate policy attention. 'One specific national policy that has important implications for the attractiveness of an economy as a production site is its investment in human resources, particularly the quality of its educational system' (Carnoy, 1996). However, such an investment in itself is not enough, since in general 'the organisation of formal education itself is increasingly out of step with the organisation of everyday life, including economic life' (Chisholm, 2000: p. 76). More specifically 'Universities in many countries are not adequately tied into a system of innovation and innovation training. This not only applies to sciences and engineering, for innovation is just as much an issue in social sciences, business practices, the law and the arts' (Carnoy, 1996:p. 90). Governments have to realise which fundamental adaptations are required of higher education institutions, and the ways in which government action can support these institutions to actually develop and implement these adaptations. The main reason for government action can be found in the importance of an adequately educated labour force for the national economy to perform satisfactorily in a more and more globally competitive environment. As is formulated sharply by Cohen (1996: pp. 146-147), 'In a world where capital moves at electronic speeds and technology leaks quickly, how can a nation stay rich and powerful if its people become dumber than those of other nations?'

While the case for an appropriate and active government role with respect to ICT and higher education can clearly be made on the basis of the above quoted arguments, it remains to be seen how governments tackle this policy issue in practice. As indicated, in this paper we will discuss the Norwegian government policy with respect to ICT and higher education. In doing so we take as a starting-point an understanding of the subtle connection between policy making, policy implementation and policy effect. We make a distinction between symbolic policy (aimed at establishing principles and attempting to unify different interest groups), comprehensive policy (an approach aimed at impacting a whole sector or system in a unified and homogeneous way), and differentiated policy (which involves making trade-offs between competing interests, on the basis of indicators or specific criteria, with the intention of addressing targeted problems or sub-sectors) (Cloete and Maassen, 2002). This distinction is based on the assumption that it is of relevance in a policy process to identify the importance of both symbols and meanings (Scott, 1995). At the start of a policy process, a policy consists of symbols in the form of words, as well as intended activities and aims that are expected to unify various interests and participants. In the policy realisation process, traditionally referred to as the policy implementation process, the actual meaning of the policy in question gradually arises in the interactions between the various participants involved. Given the fact that the participants represent various interests, the diversity of 'meanings' that arises in the interaction process is best served if the governmental agents use a flexible and differentiated policy approach.

In addition, we argue that social institutions and social policy implementers have a much greater role in change than is generally anticipated in government policy reform. The latter implies in our eyes that it is not in the first place the government policy as such that will affect the ICT development in higher education, but rather the specific circumstances at individual higher education institutions, including the opinions of the institutional leaders concerning the topic in question, that determine the actual ICT development.

4.4.1 The Norwegian governmental ICT and higher education policy

What is the position of the Norwegian government with respect to ICT and how has this position developed over the last 10 years? Let us first look at the government's interpretation of the knowledge society at the beginning of the 1990s.

The term knowledge society (in Norwegian: kunnskapssamfunnet) was used for the first time explicitly in a governmental higher education policy paper published in April 1991. In this paper the government endorses the following viewpoint: 'Without changes [in higher education] the population will be undereducated and research will be understaffed to achieve the performance the knowledge society demands' (Ministry of Church Affairs, Education and Research, 1991). Based on this viewpoint the government acknowledges that education and research will become more and more important in the lives of individual citizens as well as for the welfare of the society as a whole. Consequently the government states that its educational policy should take this acknowledgment as a starting-point. In other words, in 1991 the government explicitly intended to design policies for realising the changes needed in higher education in order to 'lift' the education and research competencies of the Norwegian society to the level needed from the perspective of a knowledge society.

In discussing the knowledge society in the policy paper the government reflects both on the knowledge dependence of Norway, amongst other things, from an economic perspective, and the consequences of the explosive growth of knowledge. With respect to the latter it is claimed that one of the most important goals for the Norwegian education system is to enable the individual citizen as well as the whole society to deal with the knowledge explosion. Emphasis is put on educating people to become responsible and critical citizens who can participate actively in Norwegian society.

57

With respect to ICT the government acknowledges that information technology is affecting Norwegian society more and more. As regards the challenges the society faces as a consequence of this, it emphasises the role of the educational system in creating an understanding of the humanistic and cultural dimensions of the ICT developments, and in helping to understand the human goals Norway wants to realise through using ICT.

This rather general vision of ICT is complemented by more specific intentions and policies with respect to distance learning and the development of an electronic knowledge-network. In this it is of importance to understand the role the Norwegian geography is playing in higher education policy making and in the practice of higher education. Being a thinly populated country with long distances between the main urban areas implies that the infrastructure for connecting the various higher education sites, as well as the whole question of access to higher education are more important in policy making than in most other European countries. As indicated in the introduction of the Norwegian higher education system, the government has emphasised the importance of distance education structures, and has seen these structures as the more or less 'natural' area for the further development of ICT in (higher) education. In Norway ICT in higher education has for a long time been identified with distance education.

Concerning the electronic network the aim was that in the period 1991-2000 all Norwegian citizens should be offered access to college or university studies all their lives, independent of where they live. Condition for realising this was the use of new technology to link all universities and colleges together in a national knowledge-network (as a part of the so-called Norgesnettet). In order for this network to get the role the government intended it was foreseen that a lot of work was needed to further develop the existing ICT services and to develop new ones.

Overall main attention in the ICT and higher education policies of the government in 1991 was given to the system level, i.e. the ICT services that were intended at the system level. With respect to the use of ICT in traditional teaching programmes the policy paper did not include more than the following general intentions. 'Both for the ordinary teaching activities and for distance learning it is of importance to develop programmes for computer supported learning, interactive videos, integration of sound, images, data, etc., in order to be able to fully exploit the possibilities ICT offers. These instruments must be pedagogically adapted, and technically developed in such a way that makes it possible to distribute them through the electronic network' (Ministry of Church Affairs, Education and Research, 1991).

In its general starting-point, this early-1990s policy approach can be regarded as an example of symbolic policy in the sense that it intended to establish principles and attempted to unify different interest groups. It offered a system level vision on ICT and Higher Education as part of a governmental response towards the demands of the knowledge society. It emphasised the humanistic and cultural role of the education system with respect to the further development and use of ICT in Norway. The 1991 policy paper provided a symbolic vision on the linking of the Norwegian higher education institutions without clearly addressing the issue of what the main purpose would be of this linking. It raised the question concerning the human goals to be achieved with ICT and the role the education system should play in this, without attempting to answer the question in other than broad and abstract terms.

The feeling one gets from reading the 1991 policy paper is that the Norwegian government at that time realised the importance of the ICT development, but it did not yet have a clear insight into nor a view of the possible consequences of this development in general, or the effects of it on higher education in particular. In addition, no reference is made to the state of the art of the actual use of ICT in the universities and colleges. It can be safely assumed that the government, at least at that time, was lacking this knowledge on the practice of ICT in higher education. This implies that the government could not develop at that time a differentiated policy addressing the problems the institutions faced when introducing and using the new technology. It also makes it impossible to develop instruments for stimulating those institutions that are lagging behind, and rewarding those institutions that are leading the ICT practice in higher education. What the government aimed at through the establishment of the electronic network was a comprehensive policy, i.e. an approach intended to impact the higher education sector as a whole.

This policy aim of creating an electronic network must in itself be regarded as a positive step towards a government ICT and Higher Education policy. It is in line with the recommendations made by some of the global experts on the issues in question later in the decade. Castells, for example, has indicated at various places that the main role of governments has become to create new connections in society, and to link people, companies, organisations, and institutions in an effective way (see for example, Castells, 1997). In that sense the electronic network is a possible example of an effective and appropriate modern government measure. However, government action in this should not be limited to connecting for the sake of connecting. Governments should preferably create networks and connections with a specific purpose, and develop an understanding of the projected output of the new networks. The latter was lacking in the case of the projected electronic higher education network in Norway. The government implicitly assumed that once the network was established innovations would take place 'naturally'. It is not clear what kind of innovations were expected, and there was no view on the support structure, including human resources and finances, needed to realize the innovations and support their use in a broader social context. In other words the first step made in 1991 can be said to be an important one, in the sense that the Norwegian government was investing in an electronic infrastructure connecting the country's higher education institutions. The subsequent steps were missing, i.e. understanding the consequences of this investment and development of instruments for supporting the higher education system in its attempts to make optimal use of the electronic structure within a politically acceptable framework.

In terms of our interpretation of symbolic policy presented above it can be argued that the Norwegian government presented the notion of a knowledge society, and the role of ICT in it, as symbols that would unify the interests of various personnel involved. The actual meaning of a knowledge society in the Norwegian context would be shaped through the interactions of those involved. Also the presentation in 1991 of ICT as an important instrument in realising the knowledge society was symbolic. The actual meaning of ICT in the practice of higher education would be 'created' in the subsequent years by the interactions of civil servants, institutional administrators, students, academic staff and others. Obviously, some of these interactions had already started in the higher education system itself before 1991. In that sense the 1991 White Paper linked to an emerging innovation in higher education. Consequently, in this early stage of the ICT and Higher Education policy process the Norwegian government could not go beyond confirming the main symbols of the new developments in the context of Norwegian higher education, especially the notion of the knowledge society. In doing so, it offered a more or less formal starting-point for a process that was expected to shape the meaning in practice of these policy symbols. The main action undertaken to stimulate this process of 'shaping meaning' was the investment in the development of an electronic network.

To summarize, in 1991 the Norwegian government's main policy intentions with respect to ICT and Higher Education were to invest in distance education, and in the establishment of an electronic network connecting all universities and colleges. The actual use of ICT in the institutions was hardly a policy issue. It was assumed to be stimulated 'naturally' through the existence of the electronic network. How is the situation ten years later?

4.4.2 The Mjøs-Committee and ICT

In 1998 a national commission was appointed to study Norwegian higher education after 2000, the so-called Mjøs Committee. This committee made a large number of policy recommendations in its report, published in 2000. These recommendations formed an important input into a governmental White Paper on higher education published March 2001. We will first discuss the Mjøs Committee's position with respect to ICT and Higher Education after which we will reflect upon the White Paper. In doing so we realise that the effects of the new government's policies in this aspect can be seen in the coming years. The 1991 policy paper created the policy framework for the ICT developments in higher education in the 1990s. In the framework of this paper the main relevance of the examination of the Mjøs Committee's position and the focus of the 2001 White Paper is with respect to the question to what extent these are in line with current ICT practice in the Norwegian universities and colleges.

The first striking characteristic of the Mjøs-Committee's position with respect to ICT is the focus on ICT as an instrument for supporting teaching and learning, as well as research. In that sense the Committee takes a different position with respect to the government's role than the government itself in 1991. Part of the reason may be related to the fact that during the 1990s, several national networks intended to stimulate, enhance and capitalise on the use of ICT in higher education had been established. The listing from the Mjøs-Committee included:

- a national working group for digital teaching and learning measures (ADL) initiated in 1997 and financially supported by the Ministry (Arbeids-gruppen for digitale læremidler);
- a continuation and further development of the Central Committee for Flexible Learning in Higher Education (SOFF) established by the Norwegian Parliament in 1990, intended to support and enhance distance-learning activities;
- the Network-university – a cooperation established in 1994 between a number of state colleges and universities trying to capitalise on the distance education market (Nettverksuniversitetet);
- the Norwegian University (Norgesuniversitetet) – a conglomerate of private companies, public sector organisations and higher education institutions with the intention to be a major participant in the lifelong learning market (main product: shorter courses and on-demand training), also financially supported by the Ministry of Education;
- the National Network for IT in Education (ITU) – a network intended to improve ICT-skills of students and teachers at all levels of the education system.

61

As illustrated, most of these networks and establishments were focussed on distance education or lifelong learning, with only one establishment having a special focus on the pedagogical content of the new technology (ADL). Given this attention for these non-traditional 'forms' of higher education (distance learning/lifelong learning), it is perhaps not so strange that the committee emphasises the importance for studies and research taking place in Norwegian universities and colleges, and not in its system level networking and connecting features. In line with this position the Committee's main recommendations with respect to ICT are focussed on the further development of ICT-aided educational provisions. Also the main role reserved for the government consists of supporting the institutions (especially financially!) to develop flexible ICT-aided educational provisions.

The Committee realises the enormous costs involved in the further development of the use of ICT in higher education. However, it does not reflect upon the consequences of these costs for the future use of ICT in Norwegian higher education. It recommends, for example, that all educational institutions from primary school to higher education should have access to broadband communications within two years. How this access is supposed to be used, which programs are necessary to train teachers at all levels to make optimal use of this access, what the costs are of the actual use of this access, including the software to be developed and distributed, and the national support structure needed, is not clear. In many ways the Committee follows a 'wishful thinking path', consisting of emphasising the relevance and importance of ICT in the day-to-day practice of teaching/learning and research in Norwegian higher education. As such, the Committee suggests that by connecting all institutions to broadband, through governmental investments in ICT aided educational provisions that the institutions should be able to use autonomously ('please, no earmarked funds in this!'), by better coordinating the current research into ICT and higher education, Norwegian higher education faces a bright ICT future.

4.4.3 The 2001 white paper on higher education

The Mjøs-Committee's recommendations formed an important input into the development of the 2001 White Paper. Like the Committee, the government emphasizes the importance of knowledge. Norway wants to be a leading knowledge nation. In order to realise this aim the universities and colleges have to adapt. Knowledge production taking place at these institutions has, for example, to become more directly linked to society's needs than is the case at the moment.

In the further development of Norway as a knowledge nation and the role of higher education as an instrument for becoming and remaining a knowledge nation, much emphasis is put on ICT. It is argued by the government that ICT is the underlying force for many of the fundamental changes in our society. As a consequence having ICT skills should be as important for individual citizens as being able to read and write. Interestingly enough in the White Paper, the government points to international research on the use of ICT in education as a source for its views on ICT and Higher Education, without pointing to some of the doubts with respect to the positive effects of ICT that have been reported internationally. ICT is seen as an instrument that supports learning. It leads to new forms of learning and examination, to new forms of organisation and cooperation, and to new roles for teachers and students. All this is regarded as positive, just like the effects of ICT on the quality of teaching and learning and students' motivation.

Referring to the Internet, it is claimed that Norway is in the forefront of the use of ICT in higher education internationally. With respect to their technological infrastructure, Norwegian higher education institutions are supposed to be far ahead internationally when compared to HEIs in other countries. What is of relevance for our discussion is the main challenge the government sees for the Norwegian institutions in this. Each institution should develop a comprehensive strategy for flexible ICT-aided educational provisions and distance learning. These are the exact words of the Mjøs Committee. What is needed for this is a good technological infrastructure, highly qualified support personal with top level competencies concerning the pedagogical use of ICT and, most importantly, regulations and an infrastructure that make it possible to attract experienced personnel, and to reward the efforts made in the development of flexible educational provisions.

The 2001 position of the government with respect to ICT and Higher Education differs from the 1991 position. Like the Mjøs Committee, the government has moved its focus from the system level to the department and individual level. Still, a large part of the policy making with respect to ICT is symbolic. 'Norway wants to be a leading knowledge country; ICT is the means through which we can be and remain a leading knowledge country; the experiences of using ICT in teaching and learning are very positive; the universities and colleges are in the frontline internationally when it comes to the use of the internet and their technological infrastructure; now the main step that is needed is for the institutions to develop an ICT strategy in order for them to make the best possible use of their advantageous position.

No mention is made of the practical problems, of the costs involved, of the training needs, and of the consequence of a further ICT development at all levels for the nature and organisation of higher education.

Referring to our interpretation of symbolic policy it can be argued that the 2001 Norwegian government is emphasising the same symbols, albeit in a different way, as the 1991 government in its presentation of the new higher education policy. An interesting question in this is the way in which the policy interactions since 1991 have actually shaped the meaning of the symbols presented in 1991, i.e. especially the notion of the knowledge society and the nature of ICT in the practice of higher education.

The 2001 Norwegian government is emphasising far more than the 1991 government the economic importance of knowledge. In 1991 the humanistic dimension was the central concern of the government with respect to the knowledge society and the use of ICT. That concern has been left behind: Norway has to be the leading knowledge country from an economic point of view. In addition, throughout the 1990s, meaning has been given to the nature and role of ICT in higher education by focussing especially on the micro-level, i.e. the actual use of ICT in teaching and learning, as well as in research. It is clear that the government is following this new meaning and has moved away from an 'ICT at the system level' perspective to an 'ICT in the classroom and in research projects' view.

Even if ICT has been given a lot of attention in political documents during the 1990s, with a special emphasis on distance education/lifelong learning, and with several networks established trying to stimulate and encourage national cooperation in these fields, one is easily struck by the little attention given to how ICT developments within higher education more generally should take place, and how potential problems should be dealt with in more operational terms. In the next section, part of this 'black box' is disclosed by presenting some evidence on the current status of the use of ICT in higher education, as seen from an institutional perspective.

4.5 Empirical Evidence on the Current Status of ICT in Norwegian Higher Education

At least on the surface, one easily gets the impression that major developments concerning the use of ICT in higher education have not taken place inside Norwegian higher education institutions. Centrally placed decision-makers in Norwegian higher education institutions indicate that teaching traditional target groups of students is still the most distinctive characteristic of the sector. Sixteen out of 20 respondents in our survey among institutional decision-makers claim that this characteristic plays a very important part in the mission formulation of their institution. Twelve out of 20 respondents also characterise their institution as a place where teaching is performed 'face-to-face', and 18 out of 20 say that such face-to-face contact is the most important aspect of 'good education'. In line with this picture only 4 out of 20 respondents think that undergraduate students have much choice with regard to options within the teaching programmes in their institution. In other words, decision-makers in higher education emphasise quite traditional features of higher education when describing current teaching practices in their own institution.

4.5.1 External factors that trigger the use of ICT

However, as indicated in the introduction, changes are taking place quite rapidly. Viewed from an institutional perspective, ICT is interpreted as an appropriate response to a number of external demands now being put on higher education.

- The need for innovation in teaching and learning is seen by several decision-makers as a prime reason for the increasing interest in ICT within their institution;
- The need for teaching and learning innovation seems closely related to the fact that 10 out of 20 decision-makers claim that national competition among higher education institutions has increased compared to five years ago and that this has strongly affected the current ICT policy at their institution;
- External policy initiatives are also mentioned as important sources of external demands. However, decision-makers claim that the sub-national or regional level of government is of equal importance in this to the national level.

Even if the data do not identify the single most important external factor behind the increasing interest in ICT, the case studies do indicate that student demand is not among the main driving forces in this. The survey does identify two important student demands, i.e. the need for more flexibility when it comes to the location of learning (9 out of 20), closely followed by the demand for more flexibility in the pace of learning (8 out of 20). However, both the survey and the case-studies indicate that student demand is not the most important reason for institutional measures regarding ICT in teaching and learning.

A characteristic trait that could be derived from the survey is that ICT is not seen as important for attracting and teaching international students. For example, 9 out of 20 decision-makers say that teaching international students is not at all important in relation to an increased use of ICT.

4.5.2 Strategic choices, objectives and leadership involvement

If one studies the underlying arguments for and objectives of the current ICT policies at Norwegian higher education institutions, 'enhancing flexibility' is identified as the central goal (17 out of 20 respondents). Second most important goal is the need to enhance the status and reputation of the institution (14 out of 20), and third come both the need to improve the quality of teaching and learning, and to create more opportunities for lifelong learners (12 out of 20). The least important reasons seem to be the need to generate institutional income, and to create more opportunities for international students.

Many decision-makers claim there exists an ICT-plan at their institution; something that could be regarded as a sign of the importance of the new technology. If one also considers the fact that the rector/president/executive board in many institutions has substantial responsibility for the development of the ICT policy (9 out of 20), including decisions on budget expenditures, it becomes clear that ICT development attracts a lot of institutional attention. The decision-makers asked in the survey also indicate that leaders from different levels in the organisation show commitment and leadership when it comes to implementing the ICT policy at their institution.

4.5.3 Implementation, collaboration and institutional measures

The outcomes of the survey indicate that the formal responsibility the institutional leadership takes in actually leading the institutional ICT policy development is combined with traditional collegial mechanisms during the implementation phase. For example, to have active standing or ad-hoc committees that regularly discuss institutional ICT issues seems to be a common way of involving people in the implementation of the policy.

Some breakthroughs when it comes to the use of ICT in teaching and learning can also be noticed. Thirteen out of 20 respondents claim that ICT is used extensively in their institution in course preparation. Also regarding the use of a web environment outside classroom activities, and with respect to the use of ICT for communication among students and instructors to support group work activities and project work, a relatively large number of respondents indicate that this has become institutional practice (8 out of 20). As a part of this picture, 13 out of 20 respondents claim that they were very satisfied with the general level of technological infrastructure in their institution.

However, there still is work to be done before one may claim that ICT is a fully integrated tool at the involved institutions. For example, the structures and incentives that promote the use and adaptation of ICT are still not in place in many of the institutions. Only in one institution does the use of ICT in education count towards promotion and tenure for the academic staff . In general, ICT-skills are seldom used in staff assessments (4 out of 20). Only 2 institutions in the sample seem to demand ICT competencies in the selection and recruitment of new staff.

The data also indicate that the implementation of ICT measures is not something the institutions in all cases take on individually. Ten out of 20 respondents say that their institution cooperates in this intensively with other national higher education institutions. And the same number of respondents also claim that multilateral relations with other Norwegian institutions are the dominating form of cooperation. Six out of 20 respondents point to bilateral cooperation structures at the national level as the dominating form, while only 3 out of 20 say that their institution either have bi- or multilateral cooperation experiences with foreign providers. As such, the data might indicate that the policy regarding the development of 'Network Norway' has at least been partly effective.

4.5.4 Implications of the strategic choices

When asked whether ICT has led to changes in the general working practices in their institution over the last two years, 11 out of 20 respondents claim that ICT has had a substantial impact. Perhaps also a change in the institutional culture and the staff attitude with respect to ICT is starting to become visible. Seven out of 20 decision-makers think that there is a positive general opinion regarding ICT on learning effectiveness among the staff in their institution.

It is striking that 18 out of 20 decision-makers see the use of ICT as essential for the strategic position of their institution. In addition, 16 out of 20 respondents claim that the use of ICT is essential for the quality of education programmes and services at their institution. The existence and use of various e-mail systems seem to dominate when it comes to type of technology used. Eleven out of 20 decision-makers asked point to this technology as most important for policy and practice at their own institution.

Data from the five institutional case studies conducted can be used to fill in this picture somewhat more. A closer analysis of the areas in which ICT is used shows, for example, that information, organisation and administrative tasks are dominating so far. ICT is, amongst other things, more used in planning courses, as a communication tool between teachers and students (e-mail), and for administrating a given study, than for direct pedagogic purposes. Thus, so far ICT has had mostly an indirect effect on teaching and learning: improving the conditions surrounding the pedagogical process.

One may argue that the policies of the national authorities that were emphasising the need to build up a comprehensive infrastructure at the system level are responsible for the ICT realities and opinions on ICT identified at the institutional level. The vagueness and lack of clear objectives on what ICT should be used for may have triggered an institutional implementation process focussing more on administrative and organisational issues than on content. On the other hand, it may also be argued that developing an effective use of ICT in teaching and learning at the grass root level takes time, implying that if such an effective use is realised it will be a more long-term effect of the governmental policies aimed at system level developments. No matter which position one takes, the existence of an ICT infrastructure is a necessity for more pedagogical uses of this technology.

4.6 Closing the Gap between Policy and Practice?

The analysis so far has suggested a mismatch between the 'ICT rhetoric' used at the national level, and the more incremental, diversified and opportunistic applications of ICT at the institutional level. Thus, there seems to be a gap between ideas concerning the 'knowledge society' at the system level and the more practical orientation that currently seems to dominate the ICT-agenda in the institutions analysed in this study. A part of the explanation of why this gap has emerged may be found in the fact that distance education and lifelong learning have been seen as the primary testing grounds for new technology in Norway: from the perspective that they are the areas where the new technology plays a very important role. While the relation between these areas and the use of ICT is indeed rather obvious, the dominant focus on this specific relation seems to have contributed to a considerable lack of focus when it comes to how ICT could be used more widely in higher education.

The attention given to developing an ICT infrastructure at the system level, while leaving the responsibility for the content and the pedagogical applications to the institutions, can be regarded as a type of policy that relates neatly to current ideas about self-regulation and increased institutional autonomy. However, such a policy can also be seen as being 'only' symbolic when the government does not take the responsibility to find out how institutions are handling the issue in practice. Therefore, one could argue that the Norwegian policy concerning ICT and higher education can be characterised as 'self-regulation with a missing link', namely, a lack of feedback through surveillance mechanisms, systematic evaluation and monitoring systems that provide the authorities with structured information on how institutions cope with the new technology, and not least, on their limitations and problems in adapting to this technology. Still, one should be very careful in interpreting 'symbolic' policy making and implementation as a process without effects.

As discussed above, the emphasis on creating a systemic ICT infrastructure seems to have been successful at the national level, with several well-functioning electronic networks having been established during the 1990s. But without personnel at the national (and the institutional) level specifying the purposes of these establishments, it is perhaps not surprising that tradition (distance education/lifelong learning) 'co-opted' the agenda.

The outcomes of our institutional-level survey also suggest that institutions may face the risk of repeating the failure of not analysing why they currently are involved in implementing ICT in their teaching and learning activities. There is currently a lot of strategic planning, leadership involvement and specific actions with respect to ICT at the central institutional level in Norwegian higher education. However, the survey and the institutional case studies both show that the institutions are not very active in actually trying to pinpoint, assess and monitor what is going on with respect to the use of ICT inside their walls. One could interpret this as a sign that symbolic aspects are not only important for national authorities, but also for the higher education institutions. One can easily get the impression that many institutions do not have a clear vision about how to use ICT, and are taking actions on the basis of more pragmatic reasons. Among the reasons specified in the survey is that many institutions seem to be afraid of being 'left behind' in the ICT-race. Showing initiatives at the institutional level is important from the perspective of the (future) strategic positioning and profiling of the institution. This picture is confirmed by analysing what decision-makers say about the future of higher education in the country. There seems to be a demand for more innovative teaching and learning schemes, more flexibility when it comes to delivery, and there is a fear that the competition in this area will increase in the years to come.

In this situation where uncertainty and complexity are dominant features, one can claim that it is very important to establish a closer tie between policy and

69

practice at the national as well as at the institutional level. The further development and application of ICT in higher education is resource demanding and making investments without having identified operational issues and how to prioritise them could be costly in more than one way. How, then, could better links between policy and practice be developed?

First, and as indicated above, there is a need for more and better information about current problems and practices in higher education when it comes to the state of the art of the use of ICT. As such, a strong case can be made for developing monitoring schemes that identify problems and inform the policy makers and other stakeholders involved on a regular and continuous basis. The underlying international study on ICT, which our Norwegian study is part of, could be seen as a first step toward a more continuous examination of measures taken and problems experienced in the area of ICT in Higher Education (see also Green 1998, NEA 2000). It is rather difficult to launch a strategy or identify measures at the national level if one does not have sufficient knowledge of what is going on either inside one's own national higher education institutions or internationally.

Second, the survey indicates that even if one can start to see new innovations in pedagogy in Norway, and new courses and innovative schemes developing, those responsible for these initiatives gain very little reward for their efforts. ICT use does not count in promotion and tenure, and is seldom an integral part of regular staff performance assessments. Our case studies in five institutions in Norway show that many of the initiatives stem from local 'entrepreneurs' taking action out of personal motivation. However, relying on the motivation of individuals is not the way to encourage systematic implementation of ICT.

The latter aspect is also important with respect to the need to change the 'culture' of Norwegian higher education. As seen in the survey, many institutions are quite traditional when it comes to their teaching and learning methods. However, many also tend to see ICT as a tool that can help them to innovate and renew their way of 'doing business'. This brings us to our third recommendation for closing the links between policy and practise: the need to better articulate what ICT can and cannot do for higher education (Wilson, 2001). As seen in the analysis of policy documents, there is (still) a tendency to identify ICT as a tool that will revolutionise higher education, not only in a Norwegian but also in an international context (see also Pedro 2001, Dokk-Holm & Stensaker, 1999). A typical expectation concerning the use of ICT in higher education institutions is, for example, that it will 'change the way a university or college does its core activities or business so that it can reach out to new needs and new targets groups' (Bates 2000: 57, see also Kirsebom, 1998). However, many universities have a history, an identity and a certain profile they (want to) hold up both for themselves and for their environment (Stensaker & Skjersli, 2002: 112). To present ICT as the tool that fundamentally alters vital functions of higher education is therefore perhaps not the best way of promoting the further use of this technology among the academic staff. The case studies in Norway, but also research conducted elsewhere (Collis, 1999; Collis & van der Wende, 1999; Falkfjell & Skjersli, 2000), show that academic staff often are reluctant to use new technology, but could change their attitude when they see, and not least, use the new technology personally and in a pedagogical context.

The implementation of ICT so far could, thus, be given two different interpretations considering the Norwegian case. In the Bates perspective, one would claim that the implementation process so far is a failure: there are few revolutions that could be identified in the Norwegian context. However, if one sees ICT as a means to renew higher education from within, opening it up through the combination of 'new' and 'old' technology for more than one way of providing teaching and learning, a more positive picture emerges. Policies should, therefore, put more emphasis on refining the language concerning ICT and adapt it to the core values in higher education (see also Wilson, 2001: p. 224).

Overall one can argue that the two policy phases with respect to higher education discussed in this paper, the first around the 1991 White Paper, the second around the 2001 White Paper, were to a large extent focussed on the concept of knowledge. However, while in 1991 the concept of the knowledge 'society' was discussed carefully with an emphasis on the many challenges Norway is facing in its transition towards a knowledge society, in 2001 the tone is different. Norway has to be a leading knowledge 'country'. While in 1991 ICT was seen as a governmental instrument to support the transition towards a knowledge society at the system level, in 2001 it is interpreted as an institutional instrument for making teaching and learning, as well as research, more effective and responsive thereby contributing to realising the goal of Norway being and remaining a leading knowledge country. In 1991 the government assumed that investing in electronic infrastructure connecting all institutions would stimulate the use of ICT in teaching and learning; in 2001 it is emphasized that investing directly in ICT as a teaching and learning, and research instrument is necessary.

One can wonder how the gap between policy rhetoric and ICT reality in higher education will affect the policy aim to be a leading knowledge country. We will not discuss this question extensively here. We do want to point at the end of our discussion to a relevant difference in approach in this between Norway and other European countries. In presenting its policy approach with respect to ICT in higher education the Norwegian Ministry of Education does not refer to a national ICT policy. The image arising out of the 2001 policy paper and the subsequent discussions in Norway, amongst other things in Parliament and in the media, is one of fragmentation. Contrary to countries such as Finland and the Netherlands, Norway does not (claim to) have a national ICT policy within which ICT in Higher Education is positioned. This lack of a national policy frame might turn out to be an additional barrier in attempts to realise the educational goals with respect to Norway as a leading knowledge country. Also in this respect the development of a monitoring system, which should include information on the way in which the ICT developments in higher education are related to ICT developments in other public sectors, as well as in the private sector, could assist the Norwegian Ministry of Education in its attempts to support the ICT developments in higher education in an effective way: the latter not only from the narrow perspective of the effectiveness of teaching and learning, and research, but also from the broader perspective of reaching the national aims and serving the national interests related to the policy aim of 'Norway as a leading knowledge country.'

References

Bates, A. W. (2000). *Managing technological change: Strategies for college and university leaders.* Jossey-Bass, San Francisco.

Brandt, E. (2001). Lifelong learning in Norwegian universities. *European Journal of Education,* vol. 36, pp. 265-276.

Carnoy, M. (1996). Multinationals in a Changing World Economy: Whither the Nation-State? In: M. Carnoy, M. Castells, S.S. Cohen and F.H. Cardoso (eds.) *The New Global Economy in the Information Age. Reflections on our changing world.* University Park, Pennsylvania: The Pennsylvania State University Press, pp. 45-97.

Carnoy, M., M. Castells, S.S. Cohen & F.H. Cardoso (1996) *The New Global Economy in the Information Age. Reflections on our changing world.* University Park, Pennsylvania: The Pennsylvania State University Press.

Castells, M. (1996). *The Rise of the Network Society.* Oxford: Blackwell Publishers.

Castells, M. (1997). *'We hebben een machine gemaakt die door niemand beheerst wordt'* Interview NRC Handelsblad, 8 November 1997.

Chisholm, L. (2000). The Educational and Social Implications of the Transition to Knowledge Societies. In: O. von der Gablentz, D. Mahncke, P.-C. Padoan, R. Picht (eds.) *Europe 2020: Adapting to a Changing World*. Baden-Baden: Nomos Verlagsgesellschaft, pp. 75-91.

Clark, B. (1983). *The higher education system*. University of California Press. Berkeley.

Cloete, N. & P. Maassen (2002). The limits of policy. In: N. Cloete, R. Fehnel, P. Maassen, T. Moja, H. Perold, T. Gibbon (eds.) Transformation in Higher Education. Global Pressures and Local Realities in South Africa. Lansdowne, SA: Juta and Company, pp. 447-491.

Cohen, S.S. (1996) Geo-Economics: Lessons from America's Mistakes. In: M. Carnoy, M. Castells, S.S. Cohen and F.H. Cardoso *The New Global Economy in the Information Age. Reflections on our Changing World*. University Park, Pennsylvania: The Pennsylvania State University Press, pp. 97-148.

Collis, B. (1999). Pedagogical perspectives on ICT use in higher education. In Collis, B. & van der Wende, M. (eds.) (1999). *The use of information and communication technology in higher education. An international orientation on trends and issues*. CHEPS, Enschede.

Collis, B. & van der Wende, M. (eds.) (1999). *The use of information and communication technology in higher education. An international orientation on trends and issues*. CHEPS, Enschede.

Dokk-Holm, E. & Stensaker, B. (1999). Kunnskapens kommersialisering. In Braa, K., Hetland, P. & Liestøl, G. (eds.) Netts@mfunn. Tano Aschehoug. Oslo.

Falkfjell, L. & Skjersli, S. (2000). Bruk av IKT som læringsverktøy i høyere utdanning. *Case-studier av Universitetet i Lund og Ålborg*, NIFU skriftserie 4/2000.

Frackman, E. (1996). Executive Management Systems for Institutional Management in Higher Education. In: *Managing Information Strategies in Higher Education*. IMHE/OECD, Paris.

Geloven, M.P.V. *et al. (1999)*. *ICT in het Hoger Onderwijs: gebruik, trends en knelpunten*. Ministerie van Onderwijs, Cultuur en Wetenschappen, Den Haag.

Green, K. C. (1998). *The 1998 campus computing survey*. (www.campuscomputing.net).

Kirsebom, B. (1998). Universiteten i it-åldern – frontlinje eller bakgård? In Bauer, M. (ed.) *Kraften ligger i det okända. Et festskrift til Stig Hagström, universitetskansler 1992-98*. Högskoleverket, Stockholm.

Ljoså, E. (2002). Credibility, legitimation and quality assurance in open and distance learning – an outline of different approaches seen from a 'back door' perspective. In Holmberg, H. (ed.) *Eight contributions on quality and flexible learning. Kristianstad. Distum.*

Ministry of Church Affairs, Education and Research (1991). *Fra visjon til virke. Om høgre utdanning*. St.meld. nr. 40 (1990-1991). Oslo, 19 April 1991.

Ministry of Church Affairs, Education and Research (2001). *Gjør din plikt – Krev din rett. Kvalitetsreform av høyere utdanning*. St.meld. nr. 27 (2000-2001). Oslo, 9 March 2001.

NEA (2000). *A survey of traditional and distance learning higher education members.* National Education Association (nea), Washington D.C.

Pedro, F. (2001). Transforming On-campus Education: promise and peril of information technology in traditional universities. *European Journal of Education*, Vol. 36, pp. 175-187.

Rekkedal, T. (2002). Quality assurance in Norwegian distance education: the background of NADE´s quality standards with reference to some European initiatives. In Holmberg, H. (ed) Eight contributions on quality and flexible learning. Kristianstad Distum.

Scott, W.R. (1995) Institutions and organizations. Thousand Oaks: Sage.

Stensaker, B. & Skjersli, S. (2002). Organising ICT-initiatives in higher education: a reflection on the critical factors. In Holmberg, H. (ed) *Eight contributions on quality and flexible learning*. Kristianstad. Distum.

Wilson, J. (2001) The technological revolution: reflections on the proper role of technology in higher education. In Altbach, P., Gumport, P. & Johnstone, D. B. (eds.) In *defense of American higher education*. Baltimore. The Johns Hopkins University Press.

5 The Creation of the Finnish Virtual University – The First Three Years

Pekka Kess, Finnish Virtual University, Finland

5.1 Introduction

Perhaps unsurpassed in the history of Finnish academia remain the success stories of Johannes Petri (1357), Petrus Roodh de Abo (1416) and Olavus Magnus (1435), the three Finnish scholars before Finland existed as a nation, who became rectors of Sorbonne in the middle ages. Ever since, the Finns have played a role in Academic Europe. In recent years, Finns seem to have been reassuming similar pivotal positions in the new eEurope. Many have become familiar with the Nokia phenomenon and their slogan, 'connecting people'. The Finnish Virtual University is happy with less (Kess, 2002). Connecting academic communities virtually would suffice.

The Finns populate a land strip of the same size as Britain, with only one-tenth of the number of people. The number of universities, however, totals 21 with 36 polytechnics. This means that units are small and scattered around. It is difficult and costly to deliver quality education throughout the country under such circumstances.

Among many strengths of the Finnish higher education (HE) there are well-equipped and networked modern campuses with nomad students virtually grazing around with mobile gadgets. Another asset is the commitment of the government and the HE institutions to the shared vision of a knowledge-based society with a national virtual university at its pivot.

Moving from the current situation into the desired direction is not an easy and straightforward job. On the contrary, the list of challenges is dauntingly long and it tends to expand when we move on. Nevertheless Finland seems to be progressing. The federal funding for the project is ensured at least for the next two years. The sustainability of the Finnish virtual university model will not be put in real mode test until 2004, when the pilot phase is over.

Finnish universities and polytechnics have made considerable investments in

ICT in recent years. At present, all of Finland's research and arts universities, most of the polytechnics and the most important research institutes: a total of about 90 organisations have joined the Finnish university network (FUNET). All higher education students can use the services provided by FUNET. During the past decade, traffic in FUNET has grown at an annual rate of 150%, or, in other words, it has doubled every nine months. This growth has been made possible by the unbelievable rate of development in the information transfer technology of telecommunications networks.

In general, universities have a relatively good and well-functioning ICT infrastructure. Teachers have nearly enough personal computers at their disposal. The greatest needs are in the area of student workstations. Another problem is insufficient technical and pedagogical support services. Information technology expertise possessed by individual students can, however, be a significant resource in the development of teaching applications. The problem often lies rather in the attitudes of teachers and even in their fear of losing authority.

According to a survey conducted, teachers mainly use information technology to prepare lectures and assignments, maintain contacts with other members of academia, acquire and process new information, and to conduct their research. Students mainly use information technology to complete their individual assignments, communicate with each other and their teachers, and acquire new information (Sinko, 2000).

The same survey showed that ICT has thus far not had a particularly profound effect on how teaching at the universities in general has been carried out. This does not mean, however, that new teaching practices have not been created through utilising these technologies. Many individual pilot projects show that when coupled with innovative pedagogical thinking, technology opens up interesting possibilities for revitalising higher education.

5.2 Rationale for Going Virtual

In many countries the fear of intensified external competition in HE or high commercial expectations have been driving forces in pushing universities towards virtualisation of their education services. This was not the case so directly in Finland. The virtual university initiative was first outlined in the national information society strategy for education and research for the years 2000-2004 adopted by the Ministry of Education and the Finnish government.

Based on the analysis of the global situation of Finnish higher education, the vision for improving higher education through going virtual was outlined and the plan for implementing it launched: by the year 2004 a high-quality, ethically and economically sustainable network-based model of organising teaching and research will have been consolidated.

A virtual university will be set up by 2004 based on a consortium of several universities, business enterprises and research institutes. It will produce and offer internationally competitive, high-standard educational services.

The aims have evolved from the original documents into this list:
– to enable networking in teaching, studying and research;
– to develop a new model of network based cross-university operation;
– to diversify university studies;
– to develop university curricula;
– to improve the quality of teaching and studying in higher education;
– to make better use of the ICT networks;
– to improve the competitiveness of Finnish academia.

5.3 Implementation

The virtual university will be established in stages. At the initial stage, the ministerial virtual university task force has coordinated the project. In connection with the negotiations on target outcomes between the ministry and the universities in spring 2000, the universities committed to establish the virtual university consortium.

The ministry of education chose then, based on applications from universities, about twenty specific inter-university projects to be funded until 2003. These projects are anticipated to play a key role in shaping the services of the virtual university and a substantial number of exemplary net-based courses and study programmes.

The development unit was set up in August 2000 to coordinate the start-up phase and emerging services. The consortium contract between the universities was signed up, the steering committee elected and the action plan approved by the consortium in early 2001. The virtual university strategy document is under preparation and will be adopted by the consortium in 2002.

All students (first degree students, postgraduates or open university students) of any member university will be eligible for studies in the virtual university or, to put it more precisely, for studies in any member university of the virtual university consortium. Students can take courses relating to their degree programmes in the virtual university, but it is the home university that will award the degree.

The main distinctive features of the FVU compared to other virtual university initiatives is the following:
- it is a national initiative that involves all of the country's universities;
- it is set in a context of a national information society strategy to improve the quality of teaching and learning at universities and offer learners greater access and flexibility through the integration of technology;
- it is not solely targeting the development and/or marketing of totally on-line courses to learners outside of its borders or responding to a competitive threat; and
- while not yet realised or fully planned, the FVU has a comprehensive vision of including teaching as well as research and support services.

An effective FVU was envisioned to offer the following advantages:
- freedom from the restrictions of time and place;
- flexible programs/courses and possibilities for individual additional studies;
- international exchange of educational material;
- cooperation and co-ordination of development work;
- saving on space and facilities;
- efficient use of time.

The FVU was also seen as an opportunity to address more general weaknesses in Finland's higher education system such as the lack of a tradition in collaboration among universities and a slow progression of studies for students. The FVU was not the result of international competitive pressure. While improvement in the competitiveness of Finnish academia is discussed, Finnish university students do not pay tuition and receive subsidies for books and living costs while studying at Finnish institutions. This and the relatively small market for programs in the Finnish language obviously limits opportunities for foreign institutions.

Due to the large expenditures required to develop virtual university programs, there was the need to pool limited resources to achieve economies of scale by connecting the work across Finnish universities. In addition to achieving economies of scale, the FVU was intended to address barriers to on-line learning applications including technical, pedagogical, social, administrative and regulatory issues. Key personnel involved with the FVU emphasize that the initiative does not require or promote full virtuality as the only model. Most people during the visit programme were very resistant to the idea of offering fully on-line alternatives.

5.4 Challenges

The demand for tertiary level graduates has been estimated to be so high in the near future that the percentage of the tertiary level degree holders from each age group entering the labour force is targeted to rise to nearly 70 percent. The figure is so high that it cannot be met merely by increasing the in-take in institutions based on traditional on-site education only. Meanwhile, the average study times in HE institutions have remained quite long, and the drop-out percentage has increased in some areas.

Increasing the share of students attending HE, increasing the number of students in certain popular fields while some areas suffer from shortage of students, calls for new pedagogical solutions. Increasingly heterogeneous student groups demand improved didactics. Going virtual and meeting the more complex needs of students is one way of trying to address these problems. They are great challenges for staff development. HE institutions are facing also serious resource problems. All these challenges have to be met with limited budgets. Earmarked extra funding covers only the piloting phase.

These facts place totally new demands on developing HE. Net-based education has been welcomed by many as, if not a panacea to all problems, bringing substantial relief or at least alternative solutions to many ailments and challenges of HE. On-line education, however, is by no means rendering easy solutions. On the other hand, there is no easy way of solving the pertinent problems HE is facing by continuing to develop campus-based education either. Face-to-face education is neither cheap, nor cost-effective. So educational policy-makers have to come to grips with the same fundamental problems of education, whether seeking solutions from the net or inside the campus.

According to our action plan, key issues to be addressed and hopefully to be turned into success factors will be:
– support to collaborative design and delivery of net-based courses;

- solving IPR issues;
- integration of different modes of instruction;
- on-line tutoring;
- virtual mobility of students;
- finding a sustainable model of operation.

These issues are currently being worked on, for instance, through a number of inter-university projects. The aim of those projects is not only to design excellent courses to be run on the net successfully. It is hoped that they will serve also as examples of good practice to be scaled up. Moreover it is hoped that they will sow the seeds of a new type of academic collaboration among professionals across institutions, faculties and research paradigms. The lack of adequate pedagogical and technical support also calls for sharing resources between institutions.

The IPR issues cannot be solved adequately within one national project, because of its global nature. The Virtual university has just issued a set of applicable contract models. In this endeavour, FVU combined efforts with other educational institutions under the umbrella of the Ministry of Education, which provided best legal assistance in drafting model contracts and will hopefully provide a certain level of on-line consultancy as well.

The Finnish virtual university is not aiming at full virtuality in the course offerings. Consequently attention will also be paid to improving current teaching and studying practices through incorporating on-line elements in any courses whenever appropriate. It will hopefully lead to a new practice optimising the use and the mix of different teaching and learning modalities in a flexible way. This gives the students an opportunity to choose from various methods of course delivery and different realisation of the learning process. Flexibility should be stretched to its limits to allow maximum personalising and customising of learning environments and teaching arrangements.

A certain level of on-line tutoring is already available to open university students. The solutions developed there provide a rich source for adopting and adapting the practice for degree students tutoring. It is believed that the tutoring services will largely be distributed among member universities and the centralised services provided by the national portal will be quite thin in the beginning. In the long run, when course design and course offerings are provided on a large scale and when information systems and databases are fully-fledged, there will be possibilities for providing more comprehensive services.

The Finnish universities are small. This means limited opportunities for students within their home campus. Regardless of high rates of student mobility, there is need for dramatically expanding virtual mobility of students. Providing students flexible opportunities to pick up courses from other universities without needing to engage in a lot of time-and-money-consuming travelling, there is enormous potential for the virtual university service provision. A task force has been collected to tackle the administrative challenges of virtual mobility with credit transfers, financial transactions, student registering, etc. Even though the virtual university initiative is in the focus of the Finnish government information society strategy, it is not run top-down. Neither is it run bottom-up. It is clearly network based and managed. All the universities are stakeholders. The activities and services are thus defined, designed and will be run by innovative and enthusiastic academic networks assisted by the ministry-owned company, CSC, which is responsible for the Finnish university network and scientific computing. The current organis-ational model will certainly need many modifications to run the virtual HE service successfully in the future. Also, it remains to be seen how soon the consortium will manage to attract non-academic partners and extend its activities beyond national borders (Curry, 2001).

5.5 Finnish Virtual University Funding and Governance

The Finnish Government, through the Finnish Ministry of Education, has committed funding to the FVU until the end of 2003. The FVU has funding of approximately 10 million Euros for the first year of its development. Approximately half of the sum has been awarded to the individual universities for their development and the other half to the 20 selected inter-university network projects. The funding in each of the two following years will be on at least the same level as the first year. Additional funding of 1 million Euros has been raised for the development of a portal. According to the Ministry of Education representative, implementation projects and services were funded based on proposals and budgets submitted by the universities or the Development Unit. Financial forecasting or costing based on existing data did not occur. The Finnish Ministry of Education intends to make the FVU a permanent programme and to consolidate the service provision developed through this initiative and the projects within the programme.

All 20 research and art universities in Finland as well as the defence academy have joined the consortium that is developing and managing the FVU. Projects, such as the scientific national electronic library FinELib and the Finnish open university, SUVI, will be closely integrated in the new initiative and collaboration is intended with the Virtual Polytechnic Initiative which will be discussed further in this report.

The task force presented its report in December 1999 and a FVU consortium of all of the participating universities was formally established to:
– develop university level net-based educational services;
– coordinate educational services offered on-line, student tutoring and the activities of research networks;
– develop course information, student achievement recording and databases;
– engage in (on-line) publishing activities;

The consortium is also an important mechanism to discuss and propose solutions to practical issues such as intellectual property rights questions, technical support required in the development work, and the development of teachers' knowledge and skills concerning on-line education.

To carry out the practical development and building work, in August 2000, a Virtual University Development Unit was established as a joint service unit for the universities. The tasks of the Development Unit, which now has a staff of about 10 professionals, include:
– development of activities and the administrative structure of the virtual university consortium;
– policy design and strategic planning regarding functions of the virtual university;
– providing support to the projects initiated by universities and the consortium;
– investigating and reporting on the activities of the virtual university, monitoring and benchmarking relevant developments in other countries and reporting about them;
– publicizing the activities of the virtual university and maintaining contacts with project partners in order to further develop joint activities;
– ensuring efficient functioning of services on a practical level;
– preparing model agreements for members (for example, agreements for partner networks, copyrights, and financial transactions).

The Development Unit works in close cooperation with the Ministry of Education's Virtual university task force and with a steering committee of a subset of consortium representatives. The Development Unit also creates and maintains contacts internationally, collecting and disseminating information on global trends in order to react quickly to changes in the environment while strengthening the operating capability of the network.

A diagram of the FVU initiative is provided in Figure 5.1. The consortium and the Development Unit is not established as a legal entity. The FVU Development Unit activities are classified as one of the network projects and coordinated by the Helsinki University of Technology, the institution that is hosting the offices of three of the FVU Development Unit staff.

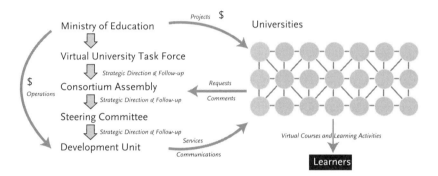

Figure 5.1 Diagram of the FVU initiative

5.6 Status of the Finnish Virtual University Project

Parallel to the work of the FVU Development Unit, at the beginning of the year, 20 three-year, joint projects between universities were initiated that will produce the first on-line courses and form the basis for the activities of the FVU. These projects include 3 regional networks, 5 joint projects aiming at providing services ('meta-projects'), and 11 networks of specific disciplines ranging from social work (SOSNET) to a graduate training programme for faculties of law. In addition, the funding provided directly to universities has been used to hire part-time or full-time support people to help in the training and support of teachers and staff who are developing on-line learning materials. The first programmes started during the academic year 2001-2002.

A virtual university portal is another major component of the FVU. The portal is intended as a functional and adaptable gateway to the virtual university net-based services with the necessary functions and services for teaching and learning including:

- course selection;
- course information including search engine;
- registration tool;
- student support services (guidance, advisory and information services, portfolio management);
- information and contact channels;
- access to national electronic library services;
- discussion forums and collaboration areas for teachers;
- support services including versatile assessment tools.

The portal will be personalised to the various users, including students, various categories of staff (teachers, researchers, administrators) and citizens. The portal will also have a tool for defining teachers' expertise profiles that can be used for defining individual, team, and organisation specific training needs. The aims of the portal, services offered, its structure, technical solutions and method of implementation were defined and were recently presented to the Consortium committee. The major anticipated challenge in the portal's implementation is the development and funding of the mid-layer of functionality, where the responsibilities of the universities and the development unit intersect. The portal was launched in the autumn of 2001 with additional functionality added by the end of 2002. The portal as currently envisaged is estimated to be fully functional in the autumn of 2004. While the FVU is intended to facilitate all academic activity including research, these services have not yet been elaborated.

5.7 Implementation Benefits and Pitfalls

While only officially launched in January 2001, the FVU has been the subject of a much longer planning process that included a task force implementation plan that was ready at the end of 1999. The FVU has already had a major impact in Finland in the following ways:

5.7.1 Building of awareness of the use of computer networks for teaching

It is apparent that the announcement of the FVU and funding awards have raised awareness and interest among university administrators, faculty, and staff in the practice of and research into the use of computer networks for learning. There is some confusion about the meaning of the terms 'virtual course' or 'virtual university' and a general belief that a face-to-face component will always remain necessary. However, the FVU has already had a beneficial impact in highlighting the opportunities provided through on-line learning and encouraging discussion and debate.

5.7.2 Putting the focus on teaching and learning

The prime goal of the FVU is to improve teaching and learning at Finnish universities. Through discussion of the use of on-line learning and teaching models, there is evidence of increased focus on teaching and learning.

5.7.3 Creating networks for collaboration, faster dissemination of best practices, and acceleration of staff capacity building

85

Each of the FVU projects that has been funded has linked faculty and staff across a number of Finnish universities. For example, one project called KASVI links the eight faculties of education in Finland. A number of projects have organised workshops and some of the workshops have attracted several hundred participants. Some of the meta-projects in particular could have a major beneficial impact. The goals of four meta-projects are summarized below:

- IT PEDA: To support universities developing strategies for virtual university activities including the creation and support of a network of pedagogical centres at participating universities. The project is also responsible for seminars for administrators on strategy development.
- Tie Vie: Training of faculty in the pedagogical use of information and communication technologies.
- IQ Form: Creating tools through which students can learn about themselves as learners and acquire skills to become more effective learners in virtual courses.

    – OVI: Project for developing the virtual environment for assessing and study counselling of students, including improving the skills and abilities required for successful studies, career counselling, and study management.

5.7.4 *Creating a forum for discussion of key issues among administrators*

The bringing together of the top administrators of all of the universities in Finland as part of the consortium has had an immediate impact with much promise for the future. The issues raised at consortium meetings range from the required support services for virtual university activities to copyright issues. Much of the topics for discussion are not restricted to virtual university matters and will impact on institutional policies for classroom teaching, such as credit transferability and copyright issues.

5.7.4 *Creating economies of scale for content development, program delivery, and critical support services*

Participation in one of the meetings of the FVU projects, the Eastern FVU project, clearly substantiated the benefits and necessity of pooling resources and efforts given the limited capacity and budgets of individual institutions. This project brings together three of Finland's smaller universities to offer jointly-designed courses, expand the course offerings, and ensure better quality and better use of local resources and expertise. There is no doubt that the FVU will face significant challenges, especially given limited resources and an operating environment that is complex and difficult to change. Several major issues and challenges in implementing the vision of a Finnish Virtual University network are identified. Some key issues and suggestions for possible solutions are as follows.

5.7.6 *Managing expectations*

The announcement of the FVU and the funding award has raised huge expectations despite the relatively limited funding to carry out the initiative and a lack of history and widespread knowledge of how to implement on-line learning in Finland. The number of full-time staff devoted to the FVU is low and success will depend on continued efforts from other specialists as well and on goodwill. There are also different expectations and different definitions of virtuality or effective virtual learning among the many participants. The impact of the FVU will probably be overestimated in the short term and, if successful, underestimated in the long-term.

To address the issue of managing expectations, the FVU needs to immediately:
- identify the various stakeholder groups including funding groups and define their expectations as quantitatively as possible;
- develop and implement a communications plan and public relations strategy that includes short overviews addressing what the FVU is and what it is not. Illustrate the funding level through comparisons with investments made in other countries;
- continue to utilize the FVU contact persons in each of the FVU participating organisations to inform academia and local interest groups and identify misconceptions and concerns.

5.7.7 Institutional commitment and strategic leadership

The level of knowledge and commitment of university administrators at the participating institutions varies. For many, on-line learning is viewed as a nice demonstration of innovation or a source of some funding to shore up weak departments. In a few cases, virtual university courses or programs are seen as an integral component of the institution's primary mission or one part of a solution set to address a challenging issue. In part, the lack of attention and funding commitment may be due to the increasingly difficult budgetary environment for Finnish universities.

87

The FVU has begun to address the requirements to increase the knowledge of university administrators with plans underway for a seminar for consortium committee members. In addition, the FVU and Ministry of Education could take the following actions:
- the Ministry of Education should propose a template for this strategic plan that identifies the critical elements that need to be addressed, such as how the administration will support the acquisition of necessary skills and knowledge by faculty, as well as indicators for evaluating progress;
- the FVU should prepare a background paper that outlines the range of organisational change strategies that have been utilised around the world and document Finnish approaches as they are developed, implemented, and evaluated.

5.7.8 Leadership on effective balance between pedagogical, administrative and ICT practices

The intention of the Ministry is to put pedagogy at the forefront of all funded and associated initiatives of the FVU. The FVU Development Unit is taking a more balanced approach putting, in addition to the pedagogical practices, the admin and ICT practices into the big picture of the FVU. There is some resistance to 'importing' models from other countries as there is a legitimate concern about adopting practices that may not be compatible with Finnish culture.

Most Finnish institutions have had some trial experiences in developing and offering on-line courses with a focus on video-conferencing classes or publishing course notes and useful course resources as opposed to implementing on-line learning activities. The departments of technical universities such as the Helsinki University of Technology and Tampere University of Technology have had previous experience with web-based offerings. The University of Helsinki Faculty of Arts, which has 9,000 students, has about 30 courses with a web-based component and approximately 5 on-line courses. The open university and continuing extension sections of the universities have also had a longer track record of offering on-line learning and have demonstrated core competencies in instructional design and learner support.

The FVU should fund and encourage projects and activities that take advantage of situations where on-line learning can add real value. A number of projects are expanding video-conference classes or providing an on-line option to a face-to-face classroom activity without examining this value-added factor or best application areas. The Ministry of Education should examine the current portfolio of projects and, working with the Development Unit, target new projects in areas that will demonstrate new pedagogical approaches that have not been utilised in the first set of projects. At the first annual reporting, it may be possible to steer some of the existing projects to use other approaches.

The FVU should develop a plan to disseminate effective practice. This plan should build on the current project meetings and presentations at the annual national conference for educators using technology and make use of the FVU web site. The web site could include a section on research evidence, strategic plan templates, discussion groups, and exemplar practices. An annual award series could recognise advances in pedagogy.

5.7.9 *Learner focus*

There is promising research that focuses on learners such as IQ FORM, and a survey has been carried out to determine and analyse the learner needs for the portal. A larger follow-up has been planned through a student and staff panel to evaluate the portal design. The FVU should consider developing a 'consumer's guide' to help students in understanding and assessing the quality of on-line learning courses and components.

5.7.10 *International strategic alliances*

With its reputation in the wireless industry, Finnish organisations receive good reception from organisations around the world. The Finnish universities are fortunate in having opportunities to link with other EU countries in research and dissemination projects such.

Given the work involved in entering into and implementing strategic alliances, the FVU Development Unit should focus on entering into a few formal arrangements each year. These arrangements could include post-doctoral and faculty exchanges, cooperation in knowledge and tool exchange in areas such as learning object repositories, and faculty and expert visits. Further information should be posted to the FVU web site in other languages with projects encouraged to provide summaries and updates for posting.

89

5.7.11 *Evaluation*

The Ministry of Education has asked each FVU project to prepare a self-evaluation report that documents approaches used and results obtained at the end of the first year. The evaluation measures are those proposed by project participants and there are no 'top-down' Ministry of Education measures. Given the importance of evaluation in managing expectations and feeding into continual improvement of the FVU's investments:

– the FVU should immediately develop an evaluation strategy for the initiative;
– the Development Unit should approach the Ministry of Education for funding a separate project for formative evaluation;
– further funding could be provided to the small evaluation project underway to develop a basket of measures and examples of evaluation instruments and templates for use by FVU project teams.

5.8 Latest Developments with Approaches and Lessons to Others

One of the basic paradigms behind the Finnish Virtual University was the idea of giving the students an opportunity to enhance their learning portfolio with courses from other universities in Finland. This has been achieved by:
– setting up an agreement to include all Finnish universities as to how the flexible studies are:
  – planned;
  – monitored;
  – accepted;
  – managed;
  – financed.
– building an information system to support these activities;
– arranging extra funding from the Ministry of Education to cover the initial extra costs of these activities;
– extensive training of students, teachers, administrative officers, etc. in the new approach.

There are many developments in the virtual university setting and abroad that can be considered innovative. The key element is the learning service to facilitate the development of on-line courses and programs by encouraging:
– learners via one-stop shopping with a wide range of information, resources and services;
– participating institutions and their faculty members with an opportunity to take advantage of economies of scale by making available a wide range of services, knowledge and resources to support the development of on-line courses and programs; and
– participating institutions with an opportunity to take advantage of synergies and economies of scale in the marketing of their on-line courses and programs at home and abroad.

References

Curry, J. (2001). *The Finnish Virtual University: Lessons and Knowledge Exchange Opportunities to Inform Pan-Canadian Plans.* Prepared for The Information Highway Advisory Branch, Industry Canada. 2001. 22 pp.

Kess P. (2002). *The Finnish Virtual University.* 4th Business Meeting in Hagenberg, Austria 28.-29.6.2002. 4 pp.

Sinko, M. (2000). ICT *in Finnish higher education: impact on lifelong learning.* Workshop on 'Application of the new information and communication technologies in lifelong learning' Catania, 6 – 8 April 2000. (http://culture.coe.fr/her/eng/catania.sinko.finland.htm).

6 ICT for Teaching and Learning: Strategy or Serendipity? – The Changing Landscape in Ireland

Jim Devine, Institute of Art, Design & Technology, Ireland

6.1 Introduction

It is a daunting task to mirror at a point in time and in a particular country the status, trends and issues concerning deployment of ICT in higher education. Indeed, the mirror metaphor is inadequate; a prism with its capacity for refraction and differentiation may be more appropriate. We can indeed ask objective questions concerning what is actually happening, infrastructure, pilot projects or institutionally-led initiatives. Of greater interest, however, are questions concerning the drivers of change and innovation and how ICT is perceived as an agent of that change. For example, are we more concerned with the possibilities afforded by ICT to better manage and control familiar environments than with the transformational possibilities that arise from consideration of the nature of teaching and learning in the various disciplines? The former is more in evidence in Ireland, but this paper seeks to put both in relative perspective. It is an appropriate time to frame such questions, as a significant debate is currently underway within Irish higher education, with the goal of identifying strategic directions.

6.2 Higher Education in Ireland

Before considering the deployment of ICT higher education in Ireland, it is useful to set the scene and to provide a contextual overview of a rapidly changing higher education landscape.

In all, more than 60 recognised institutions provide accredited higher education programmes in Ireland, including a number of small specialist colleges, which are typically affiliated to universities or which operate on a private basis. The majority of students are however located within one of 7 universities, 13 Institutes of Technology (polytechnics) or the Dublin Institute of Technology (DIT). While Ireland may appear to have maintained the binary divide, there is considerable overlap between the university and the polytechnic sectors; both offer undergraduate, masters and doctoral level studies. Funding and accreditation arrangements differ however. Universities derive their state funding through the Higher Education Authority (HEA) while Institutes of Technology and DIT are currently directly funded through the Ministry of Education & Science. Accreditation is an internal matter for universities. The recently formed Higher Education and Training Awards Council (HETAC) is the accrediting body for Institutes of Technology. DIT also has powers of accreditation. This complex arrangement is now monitored by the recently established National Qualifications Authority of Ireland (NQAI). The 1990s will be remembered in Ireland as the period during which the legislative framework for higher education was intensely debated and significant new legislation has been enacted in this period.

More than 125,000 students are registered for full-time courses of study annually, out of a population of some 3.7 million inhabitants. The participation rate of the school leaving age cohort places Ireland towards the upper end of Trow's scale defining mass higher education. However, very significant demographic change is already occurring in Ireland, with a dramatic decline in the school leaver population, a phenomenon that was encountered in the UK more than ten years ago. This change is focussing strategic debate; without radical reconsideration of how they operate and without the introduction of greater flexibility and the opening of colleges to a wider audience, a number of colleges face major rationalisation or, worse, closure or assimilation. Opportunities for participation in part-time, modular study are relatively limited. Equally, the opportunities for study by distance learning, provided by OSCAIL (the National Distance Education Centre) cannot address the full demand for this mode of learning and many adults choose to avail themselves of opportunities offered by the UK Open University and other emerging global providers.

6.3 ICT: One Among Many Issues in Higher Education in Ireland

A strategic review carried out by Skillbeck (2001), identified six major challenges facing the university sector in Ireland, which by extension may be applied also to the polytechnics (Skillbeck, p.25):

- continuing increases in individual and social demand for access to study at all levels;
- insistence by governments and employers on more economically and socially responsive education and research;
- pressure to improve quality and achieve higher overall standards;
- changing fiscal policies and priorities for public expenditure;
- requirements to improve efficiency and raise productivity;
- a progressive shift from formal, institution-bound teaching to technology-facilitated learning.

It is interesting to note Skillbeck's perception of 'technology-facilitated learning' and the nature of the issues which increased ICT deployment might address. Skillbeck goes on to assert that

> ...[U]nless the established, public sector institutions are able to achieve greater openness and flexibility they will be challenged by a variety of alternatives... including for-profit private universities taking advantage of more flexible arrangements for awards and recognition of learning; and the technology-driven 'virtual universities' (p.76).

Skillbeck's review has unleashed predictable reactions; nonetheless it presents a stark view of the higher education landscape, one which is no longer secure in itself or immune from international and commercial influences. His views, which are currently influential in shaping strategic debate, are highly cautionary in relation to the university-led initiatives in deploying ICT for teaching and learning. He cautions that

> ...only through more collaboration among universities and with media organisations can wasteful duplication and variable quality of courses and course materials and resources be avoided (p.85).

He recognises

> ...new opportunities for creative and innovative teaching and new relationships both with students and the shifting world of knowledge (p.89)

but then asks

> ...are staff motivated and adequately prepared to take advantage of the opportunities? (p89).

Since the publication of Skillbeck's report, the Presidents of Irish Universities have been engaged in a strategic exercise to map the issues he has identified. The Directors of the Institutes of Technology have been engaged in a parallel exercise. The University sector has identified 'leadership in the use of ICT in education' as one of twelve sectoral issues. Interestingly, the same issue has been articulated by the combined Institutes of Technology among eleven somewhat overlapping issues for that sector as 'flexibility and responsiveness (eLearning and ICT): teaching and learning strategies'[1]. What is clearly evident in this thinking is a systemic perspective where ICT is seen as being central to underpinning a re-engineering of familiar systems and processes. For example, no-one can dispute the improved library and research environment that ICT now facilitate, with instantaneous access to on-line journals and resources. Furthermore, the use of web-based platforms (e.g. WebCT, Blackboard) as 'Learner Management Systems' is gaining in acceptance, but again the emphasis is on underpinning traditional methods of teaching and learning in the belief that efficiency or productivity gains may be achieved and that flexible modes of provision may ultimately be more easily facilitated. Another basis that can be argued for this type of systemic approach is that it potentially increases transparency at a time when ever more rigorous academic quality assurance procedures and accountability are being demanded.

6.4 Benchmarking ICT Developments

CRE (now EUA, the Association of European Universities) has maintained a special interest group in the domain of strategic planning for ICT in higher education since 1995. Its first publication (CRE, 1996), seeking to guide universities on how to formulate strategy for ICT for teaching and learning, cautioned that at that time:

> ...only some [of the] universities ... linked their activities in the area of new learning technologies to their overall strategic objectives (p.6)

and that

> ...there is little motivation for an academic to get involved in a process for which there is little reward. Not only would research work be better perceived, there is not much in the way of support staff for course design. The negative attitudes of university administrators were also mentioned (p.6).

It is not difficult to find evidence that we have moved on significantly since 1996; indeed, there is no University or Institute of Technology in Ireland that does not make reference in its strategic plan to the potential of e-Learning as an agent of change and innovation in terms of how teaching is organised and delivered. The evidence of pilot projects that have been undertaken almost universally at this stage in the university and polytechnic sectors suggests that awareness of new learning technologies compares favourably with international trends. ILTA, the Irish Learning Technology Association, represents academics in both sectors and attracts a large audience to its annual conference; however, the scope of papers presented is concerned largely with an agenda of experimentation rather than one of major institutionally-led strategic initiatives.

Returning to the subject four years later (CRE 2000), two key issues dominated: the

> ...need for explicit strategy formulation in relation to ICT in teaching and lear-
> ning, articulated within the context of overall ICT

and

> ...the importance of strategic alliances (university consortia or university-
> industry) in underpinning university strategies (p.25).

In these concepts we find the point of departure, the divergence between institutions that are still at an experimental stage and those that have engaged with serious strategic intent in considering the potential for radical innovation afforded by e-Learning when deployed within a well-developed and robust technical and pedagogical infrastructure.

In the light of the above and in order to provide a framework for examining the evidence from Ireland, the following indicators have been considered:
- strategic policy formulation and/or initiatives
  (a) at the national or regional level;
  (b) at the level of the individual institution, university or polytechnic;
- status of the technological infrastructure: robustness, reliability and accessibility
  of the required technology;
- clear tactical and pedagogical reasons at the level of the faculty or department for the re-engineering of courses;

–   resolution of staff development and reward issues: clear rationale for
    investment of effort by career-oriented academics in innovative actions
    aimed at improving teaching and learning;
–   recognition of the multimedia and instructional design support needed to
    enable the individual academic or course group to migrate courses to a
    technology-supported environment;
–   positive disposition of staff to innovations involving ICT;
–   similarly positive disposition on the part of students;
–   capacity and appetite for organisational change and renewal;
–   nature of the physical spaces provided on campus.

These benchmarks will be considered in the next section.

6.5 Examining the Evidence

6.5.1 Strategic policy formation

A feature of higher education policy in Ireland has been the absence of
intrusive interventions on the part of the Ministry of Education & Science or
the Higher Education Authority. The growth of the higher education sector
during the period of rapid expansion in the 1980s and 1990s came about in a
climate where demand for places far outstripped the capacity of the system to
provide them. However, this laissez faire approach has resulted in a degree of
academic drift with the result that in Ireland's rapidly growing economy, key
labour market skills are in short supply and at the same time some disciplines,
notably in the sciences, are proving unattractive to new entrants. Clearly,
changing demographics are forcing a re-think. A side effect of the laissez faire
approach is the absence at national level of any strategic planning or strategic
enabling initiatives in the field of ICT for teaching and learning. Initiatives such
as the CTI Centres and the TLTP programme in the UK and the consequent
emergence of specialist centres of subject-matter expertise are entirely absent
in the Irish context. However, links at faculty and departmental levels between
Irish universities and the specialist centres in the UK are not uncommon. The
Higher Education Authority does invite tenders from universities for funds it
makes available for innovation involving application of ICT in teaching and
learning[2], but no similar scheme is available within the polytechnic sector.
Individual institutions have responded in a strategic manner to a greater or
lesser extent. Experimentation with web-based support platforms is universal,

although in a majority of cases it is targeted at campus-based students as a 'value-added' support. In the majority of cases, these platforms are used to manage the learning environment, e.g. to provide essential course materials (largely text-based or PowerPoint presentations), bulletin board facilities and a modicum of class discussion opportunities. Ironically, growth in interest and in the acceptance of 'Learner Management Systems' appears to be accompanied by a decline in interest in the development or deployment of innovative curricular resources, e.g. those involving modelling and simulation. It would appear that the prevailing view is that such developments are largely outside the scope of what an individual academic can achieve and that they are best left to private sector multimedia and e-Learning developers. This approach can come home to roost with the high costs of licensing such materials from private providers.

6.5.2 *Technological infrastructure*

Attempts to exploit the advantages of ICT in terms of providing enhanced flexibility are largely dependent on the status of the technological infrastructure. Staff and students must have convenient and reliable access to a robust ICT infrastructure, preferably supporting broadband, nationally and locally. This is the sine qua non of confidence building. Ireland fares reasonably well at this time, at least at the level of the backbone on and between the major university and polytechnic campuses. The network to support higher education is provided by HEANet and comprises a backbone running at 155Mbps, although this is being currently upgraded. External links include the following:

- INEX (Irish neutral exchange) 40Mbp;
- UKERNA (UK NREN) 155Mbps;
- Géant (pan-European R&E network) 2 x 155Mbps (upgrade to 2.5Gbps in Oct 2002);
- Abilene (US R&E network) 155Mbps.

There is a high level of technical and commercial redundancy between all circuits, and failure of one or more will cause traffic to re-route over the other circuits[3]. While this provides an impressive backbone, access by students whether on campus or off campus remains uneven. A recent survey conducted by the Union of Students in Ireland highlighted the difficulties often experienced by students seeking to access basic computing facilities in the crowded computer laboratories and libraries of their respective institutions.

99

While many students [and academic staff] now enjoy remote access to campus networks, this is almost exclusively at low access speeds; ADSL, cable modems and similar technologies are only now becoming available to Irish subscribers and at costs that are exorbitant when compared with those that exist elsewhere in the EU. The strategic enabling factor: accessible broadband technology has yet to make its impact in late 2002.

6.5.3 Tactical and pedagogical reasons for innovation

The changing profile of the student population and pressures for flexible access to higher education are the key drivers for what might be regarded as process innovation. Recent legislation, applicable to the Institutes of Technology, for example, explicitly champions the 'learner'[4]. There is no significant experience base within the university and polytechnic sectors in conventional open and distance learning provision, much less in the development and support of on-line learning. Nevertheless, considerable interest in flexible eLearning provision is evident as institutions seek to come to terms with new realities. Again, however, it must be noted that the emphasis would appear to be placed on finding new ways to manage familiar processes that are in danger of becoming progressively more dysfunctional in a mass higher education system. Increased diversity in student intake and pressures on students to work while studying place new and different demands on academics. Well-designed ICT learning environments are widely recognised as having the potential to augment and in some cases replace what have been up to now exclusively face-to-face interchanges. Pilot projects have been widely undertaken, but published results of evaluations, if they took place at all, are conspicuous by their absence. One exception is the study undertaken by Fox and MacKeogh (2001) which provides valuable insights into the experiences of students studying a module in the humanities as part of the OSCAIL distance education undergraduate programme.

6.5.4 Staff development and reward structures

The balance between teaching and research and its influence on prospects for career progression have long been recognised as key factors in determining academics' investment of effort in pedagogical innovation and in the development and deployment of learning technologies. The need for balance has been recognised in the UK through the much criticised parallel assessments of research and teaching. Similar evaluations do not occur in Ireland at this time. Indeed, unprecedented levels of funding are now being made available for research under the National Development Plan and this aspect of academic life is clearly in the ascendant. Evidence of a renewed focus on teaching and learning can however be found in the creation of posts of 'Dean of Teaching' at Vice Presidential level in a number of universities. The creation of such posts signals a new strategic intent and the consequent focus in the universities on methodologies for teaching and assessment and in particular on the role of enabling technologies nurtures a culture of innovation and a demand for staff development and renewal. The very fact that a recognisable infrastructure has been provided encourages teaching/learning projects of a more ambitious nature. The creation of learning resources ('content') to a much higher production standard becomes achievable. Innovations can be seen to target areas such as active learning, provision of greater flexibility (time/place independence) and learning supports for a more diverse student body. Conversely, within the polytechnic sector, contracts of employment for academics are highly prescriptive in terms of teaching loads and while one might expect a higher level of interest in pedagogical innovation, this is not evident to date, indicative perhaps of a lack of targeted funding. Furthermore, no specific provision is made for 'Dean' posts comparable to those in the university sector and responsibility for pedagogical development and innovation rests squarely with Faculty and Departmental Heads.

One point on which there is unanimous agreement is the need for improved staff development opportunities focussed on the academic as teacher, facilitator and mentor. There is an almost universal emphasis on the ability to use advanced audiovisual resources (e.g. PowerPoint presentations with varying degrees of multimedia complexity) whether or not these are also integrated into web-based environments for on-line delivery. The inauguration of the Institute for Learning and Teaching in the UK and its current development trajectory is being followed with interest in Ireland, where to date no steps have been taken to formalise the academic profession in this way.

6.5.5 Support for course resource development

The rapid rise in popularity of out-of-the-box e-Learning platforms, e.g. WebCT, may ultimately have a negative side effect. The relative ease with which basic course resources can be integrated within such environments and the focus on learner management features tends to offset demands for the development of high production value multimedia and e-Learning resources.

The multidisciplinary teams that are required for quality courseware production and the high cost and relatively long lead times make the 'quick and dirty' solution more attractive if ultimately more limiting.

The university sector has addressed this issue through the establishment of 'Learning Development' centres or equivalent. The success of such centres and the degree to which their services are in demand is variable; it is as much conditioned by institutional strategic priorities as by the availability of budget and staff resources. In the polytechnic sector, scope for the establishment of such centres is extremely limited and in one case where a centre has been set up, its survival is critically dependent on revenue from external projects.

The strong discipline-based support that has developed in the UK is simply not available in Ireland; the scale of operation does not appear to warrant it. Therefore, from an Irish perspective, collaborative approaches should underpin developments and these should include universities in other countries, commercial organisations and private publishers.

6.5.6 Disposition of staff and students

The dearth of published evaluations makes it difficult to form a definitive view of student perceptions of e-Learning. Anecdotal evidence suggests that where a real need is being addressed, e.g. inability to attend on campus on a full-time basis, students and staff are willing to avail themselves of on-line opportunities, however imperfectly these are developed and supported. Students frequently seek efficiency gains in terms of time on task and are not always interested in exploring 'rich learning environments' if this makes further demands on their time. The risk of inculcating habits of surface as distinct from deep learning cannot be discounted.

Staff require reassurance that their workload will not increase in an uncontrolled manner when large groups of students enjoy greater access to them through e-mail or on-line discussion groups.

Student feedback suggests that they particularly like the bulletin board and transactional features of on-line course environments. This clearly mirrors the move to web-based enquiries and transactions in all facets of life.

The prevailing disposition is towards layering new learning technologies onto conventional course structures, an incremental and relatively low risk strategy. It is recognised that fully on-line course delivery requires a significantly different pedagogical model and is intensely demanding in terms of the high production values for course resources. It is not surprising therefore that such models are slow to emerge. In the case of OSCAIL, the National Distance Education Centre, an on-line programme at M.Sc. level is now offered, with a student body largely drawn from its own undergraduate programme, which in turn is offered primarily through conventional text-based distance learning.

6.5.7 Capacity for organisational change and renewal

There are compelling reasons for strategic change in higher education in Ireland. A number of initiatives relating to e-Learning have already commenced, demonstrating a capacity for change and innovation within the higher education sector.

The Royal College of Surgeons of Ireland (RCSI) has formed a strategic relationship with a commercial e-Learning company to produce on-line courses for a global market. Dublin City University, with the support of a wealthy Irish entrepreneur and industrialist, has created the EEOLAS[5] Institute, located at the Citywest Digital Campus near Dublin. This initiative also includes strategic alliances with several US universities. The stated goal, recently reported in *Education* (2002), is to develop.

> ...a variety of research and educational initiatives that tend to prosper in an environment of enterprise, where university, business and industrial leaders freely interact and collaborate.

The development of what is being described as a 'Virtual Academy' as an integral element of the 'Digital Hub'[6] project further indicates tentative steps that are being taken to achieve collaboration across higher education institutions in an emerging global marketplace.

6.5.8 Nature of the physical learning space

In thinking about the potential benefits of ICT in teaching and learning, there is a tendency to lose sight of the physical environment in which students and staff go about their daily business. The image of the student engaged in scholarly pursuit in front of a terminal in the crowded spaces of computer laboratories or in library buildings built for another age somehow does not quite fit. Opportunities to design or refurbish buildings for the networked e-Learner arise infrequently, but in the Irish context a number of major capital projects have been realised in recent years, notably new library buildings for universities and polytechnics. Re-conceptualisation of the library as a networked, media-rich learning and information resource centre, perhaps more than any other architectural realisation, signifies the new genius loci.

A further departure, again indicative of trends in faculty development, is the opening in September 2002 of the Quinn School of Business at University College Dublin. The new building which houses the School has been designed with a specific pedagogical approach in mind. Undergraduates will be required to have a notebook computer from the outset and will be expected to pursue their studies in an e-Learning environment both on and off campus. In its layout and in the networked services provided, the building uniquely responds to a new learning paradigm.

6.6 Inferences and Conclusion

Arguably, educational goals at university level fall into three broad categories:
- knowledge and skills acquisition; competence with tools and techniques;
- socialisation into the canons of particular disciplines or professions (Brufee, 1993);
- development of intentional learning: making learning the goal; developing the individual as a self-organised learner; fostering critical thinking, reflective practice and active open-ended enquiry (Scardamalia et al., 1996).

To what extent is there evidence that the deployment of ICT to support teaching and learning supports these goals in a balanced way? The situation in Irish higher education as of 2002 suggests that, while much progress has been made, it is limited in scope and largely supports the first goal. Are we in danger of losing sight of exciting possibilities for pedagogical innovation, that are dependent on an understanding of technology in what Salomon et al. (1991) refer to as the 'person-plus' mode, the personal computer (and now handheld devices) as prosthetic, enabling the learner to extend their learning, experimental and computational horizons?

In summary, the situation in respect of ICT in higher education in Ireland is as follows:

- there is a strong sense of awareness of the potential of e-Learning, at least in terms of achievable low level goals and there is universal evidence of pilot implementation projects;
- published strategic plans of all major HE institutions address learning technologies and e-Learning;
- stand-alone multimedia resources are widely used at Departmental /Faculty level, e.g. CD-ROM, multimedia learning resources and these are also widely available in academic libraries;
- guiding students to Internet resources and the use, for example, of book publishers' accompanying web-based resources, is common practice, particularly in science and business disciplines: this is a viable alternative to creating 'in house' resources;
- no strategic e-Learning initiative has yet been taken at national level, but consideration is being given to such an undertaking at this time. The Higher Education Authority is expected to issue a call for expressions of interest in Autumn 2002;
- the backbone infrastructure is adequate in terms of supporting reliable access to on-line courses and resources; however, access from the home is limited by the unavailability of broadband services and the relatively high cost of dial-up internet access;
- strategic planning for organisational change is already taking place at national level within the university and polytechnic sectors and e-Learning is recognised as an important element in a changing landscape;
- collaborative ventures with universities in other countries, particularly in the US, are being explored;
- growth in the licensing and deployment of 'Learner Management Systems' and extensive piloting of such applications are leading to an improved understanding of the on-line learning environment. However, there is a danger of sub-optimal achievement if such systems are used solely for information provision and for relatively low level communications tasks;
- it is recognised that the development and implementation of high production value courseware requires collaboration of an extensive nature; evidence to date of such collaboration is not to be found in the national context;
- there is a growing body of expertise, arising from the many pilot projects that have been undertaken; harnessing this expertise to define and achieve strategic goals is now an important priority.

Higher education in Ireland is entering a period of transformation. Participation rates are high and the profile and demands of the student body are rapidly diversifying. In attempting to frame a strategic response, universities and polytechnics recognise that e-Learning is a key enabler of change. The status of knowledge and experience of ICT deployment compares favourably with the most highly developed nations. What has been achieved to date is largely the result of the efforts of higher education institutions acting independently. To take the next step will require strategic collaboration, the models for which are currently embryonic and ill-defined.

References

Brufee, K. (1993). *Collaborative learning: Higher education, interdependence and the authority of knowledge.* Baltimore: John Hopkins University Press.

CRE, (1996). *Restructuring the University – Universities and the Challenge of New Technologies.* CRE doc N°1. Geneva, CRE

CRE (2000). *Formative Evaluation of University Strategy for New Technologies in Teaching and Learning.* cre doc N°5. Geneva, CRE.

Editorial (2002). New University Institute Announced at Citywest Business Campus. *Education,* Dublin, Volume 15, No. 2, pp.3-4

Fox, Seamus & MacKeogh, Kay(2001). The Picture eLearning Project: On-line Resources; Pedagogical Techniques; Higher Order Learning and Tutor Support. OSCAIL, Dublin City University.

Salomon G., Perkins D. & Globerson T.(1991). *Partners in cognition: Extending human intelligence with intelligent technologies.* Educational Researcher, 20(3), pp.2-9.

Scardamalia M,. Bereiter C. & Lamon M.(1996). The CSILE project, in McGilly K. (ed.) *Classroom lessons: Integrating cognitive theory and classroom practice.* Boston, mit Press.

Skillbeck, Malcolm (2001). *The University Challenged – A Review of International Trends and Issues with Particular Reference to Ireland.* Dublin: Higher Education Authority and Committee of Heads of Irish Universities.

Notes

1   Working papers, April 2002, available to the author as participant

2   See HEA web site, www.hea.ie

3   Source: M.Norris, Senior Technical Officer, HEANet.

4   Qualification (Education and Training) Act, 1999

5   'eolas', translated from Irish means 'knowledge'
6   The Digital Hub [see www.thedigitalhub.com] is a major Government funded pro-
    ject intended to underpin renewal through the creation of a digital media district
    comprising R&D facilities [Media Lab Europe as the anchor tenant], enterprises and
    educational programmes.

7 Learning Technology in Higher Education in the UK: Trends, Drivers and Strategies

Nick Hammond, University of York, United Kingdom

7.1 Introduction

For the great majority of students now entering higher education, it would be almost inconceivable to carry out everyday communication, information-seeking and many other tasks without the use of interactive technologies. Use of the web, e-mail, advanced mobile phones, game consoles and a variety of work and entertainment applications sets the context for much of everyday life for young people. The use of information and communication technologies (ICT) is rapidly becoming the mainstream for communication and information access rather than a quaint nerdish backwater, and this growth can only accelerate over the next decade. Students will increasingly assume that Higher Education, as a primary broker in the exchange of information and knowledge, should be at the heart of this electronic culture. The question for higher education policy is not so much how technology should be factored into existing teaching and learning processes but rather how the Higher Education system as a whole should adapt to this changing context and the changing set of expectations in which it finds itself.

Over the past decade, the UK higher education system has become increasingly aware of the importance of responding to these changes, and has developed a variety of policies, programmes and support for ICT development and use. It would not be feasible within the scope of this Chapter to describe all of these initiatives and their impact: instead I will review some of the key drivers of ICT use within UK HE, some enablers and barriers to change, and consider some features of current usage and prospects for future development.

Figure 7.1 provides an overview framework of influences on the use of learning technologies (LT) in higher education. The higher education system exists within a political, cultural and social context which influences educational policy and practice in a variety of ways. The most powerful influences at this broad level include: (a) a range of higher education policy drivers, (b) developments in information and communication technologies of relevance to education and (c) the beliefs and expectations of society concerning the purposes and nature of higher education in relation to wider concerns (whether on the part of students, employers, the media or taxpayers more generally). These are illustrated in the outer ellipse of Figure 7.1. The three groups of factors (policy drivers, technology development and beliefs about education) will not only be interdependent but will also impinge on the use of learning technologies, both directly (the central ellipse) and indirectly, via various funding and support agencies and bodies (middle ellipse).

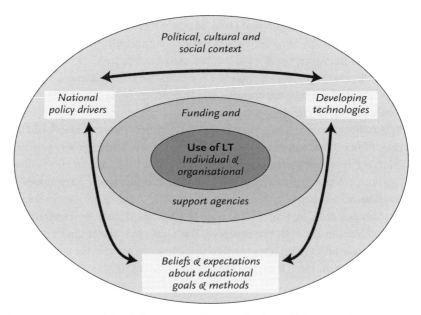

Figure 7.1 Framework for influences on the use of LT in UK higher education

The next section discusses some of the key drivers of ICT use in UK higher education. However, the path from policy to practice may be an uncertain one, and the following sections discuss aspects of current usage and some of the barriers and enablers of change. Finally, we look at some future trends.

7.2 Drivers of ICT Use in UK Higher Education

So what are the key drivers in the UK which influence the development and take-up of learning technologies? We should note from the start that some drivers might be expected to impact fairly directly on LT use (for instance a reduction in the price of PCs, or a funding policy towards increased network provision) whilst for others the impact will be indirect (for instance a change in the law relating to provision for disable students, a change in student funding resulting in a greater need for flexible working). Furthermore, the effects of some drivers on the actual usage of learning technologies by learners or teachers may be hard to predict and indeed may have unintended outcomes: it would be surprising if this were not so in view of the complexity of HE systems and mediating variables, and the difficulties of modelling systemic change.

A couple of examples illustrate this point. In 1992, the UK higher education Funding Councils set up the Teaching and Learning Technology Programme (TLTP) with the intention of developing a large amount of computer-based materials (generally termed courseware in the early 1990s) for use in HE. Analysis had identified a lack of courseware as a major block to the exploitation of learning technologies, and TLTP was conceived with the objective 'to make teaching and learning more productive and efficient by harnessing modern technology'. While the programme did develop much high-quality courseware (commended in an initial evaluation study: HEFCE, 1996) and a later evaluation study (Haywood et al., 1998) showed moderate take-up of the materials (by about 33% of the departments sampled), it proved impossible to link take-up to productivity or efficiency gains. Both evaluations pointed to the considerable impact on institutions and staff involved in the projects, and to the raised profile and credence of work in the learning technology area. So whilst the initial aim of increased productivity may not have been met (and certainly was largely unmeasurable), the funders could take comfort from the unexpected outcomes in terms of culture change.

A more recent example comes from analyses of the impact of the Teaching Quality Assessment (TQA) conducted by the Quality Assurance Agency (QAA), whose current mission is to 'promote public confidence that quality of provision and standards of awards in higher education are being safeguarded and enhanced.' Nearly all departments in HE institutions were reviewed during the period 1995 to 2001, and for this review the focus was more on 'quality assessment' (and in particular on providing detailed and tangible evidence of quality of provision and of quality management mechanisms) that on 'quality

111

enhancement'. (The balance is somewhat changed in the current round.) This emphasis on robust measurement of 'quality indicators' led a number of observers to comment on the potentially stifling effects of the TQA exercise on innovation: certainly some departments took the view that well-documented and well-tried traditional methods would earn more 'QAA points' than more innovative and potentially risky methods, which should not be adopted until the TQA exercise was safely over, if at all. At a more pragmatic level, the rigorous quality control mechanisms for course review adopted by many institutions (perhaps involving much documentation and pondering by committees) have made it harder and slower for the individual to introduce innovations into their teaching. These examples do not necessarily imply that the TQA exercise was counterproductive; the point I wish to emphasise is that the impact of such exercises on the adoption of ICT may well be mixed.

I will now consider the current main drivers of ICT use in higher education. Brown, Davies, Franklin and Smith (2002) have recently identified 12 key drivers for the use of ICT in UK Higher Education. These are listed in Table 7.1 with a brief explanation of the UK context where necessary.

A number of these drivers relate to explicit government policy: widening participation and diversity, employability, and quality assurance and enhancement. It is of interest that none of these current policy drivers relate directly to ICT use although all have ICT implications. This contrasts with the more explicit policies in the early 1990s relating technology use to potential productivity gains and which led directly to TLTP and other programmes. Our current government has committed to participation rate in HE of '50% of those aged 18-30 by the end of the decade, while maintaining standards', to 'make significant, year-on-year progress towards fair access', to 'bear down on rates of non-completion' and to 'strengthen research and teaching excellence'. Skills for employability is a government concern for at least two reasons. First, it is important to its widening participation strategy because if it succeeds there will be more graduates looking for jobs. Second, the government believes that a good supply of highly-skilled employable graduates is essential for national economic and social well-being.

Table 7.1 Twelve key drivers of ICT take-up within UK higher education (adapted from Brown *et al.*, 2002)

Driver		Comment
1	Widening access and student diversity	*Government targets for widening participation in HE, and related policy for fair access to HE; legal requirements for student support in relation to Special Educational Needs and Disability Act (Sept 2002)*
2	Employability	*Two underlying policy drivers: increased participation in HE places greater emphasis on graduate employment skills; linkages between employment and economic and social benefits*
3	Quality and standards	*Requirement for quality assurance monitoring and reporting of standards, through the Quality Assurance Agency (QAA); explicit links to quality enhancement.*
4	Increased IT literacy of students	
5	Student expectations of ICT use	
6	The earner-learner	*Exigencies of current student funding require many 'full-time' students to take part-time employment; need for increased provision for mature, part-time students in employment*
7	Increased provision of part-time courses	
8	Globalisation of learning	
9	Professionalisation of teaching	*The Institute for Learning and Teaching in HE (ILT) provides accreditation HE teachers, and new lecturers are normally required to take an induction programme.*
10	Staff shortages in key areas	
11	Staff handling larger groups	*Increased participation without concomitant increases in staffing resources results in larger classes*
12	Increased it literacy of new staff	

Other drivers listed by Brown *et al.* (2002) can be seen as consequences of broader policy or provision. The increase in 'earner-learners' (that is, students needing part-time employment to make ends meet) results partly from the greater inclusivity of widening participation and in part from policy on student funding; the demand for more part-time courses is also partly a result of these pressures. The move towards the professionalisation of teaching in higher education can also be viewed as a result of greater government concern for accountability and quality standards. The increased IT literacy in students and new staff, and the higher expectations of the role of ICT in higher education can be categorised under the 'beliefs and expectations' in Table 7.1, whilst the remaining drivers (globalisation and staff shortages) can be viewed as relating to changes in the wider political, social and cultural context.

An obvious omission from Brown *et al.*'s list is the set of drivers relating to technology development itself: the reducing cost of access to IT and the ever-increasing range of technologies and their potential applications of ICT within education. Whilst particular technology developments can be seen also as enablers of change, in the exploratory phase of adopting a technology (for example in the adoption of e-mail or the Web into educational practice), successful use by early adopters is perhaps more accurately categorised as driving rather than merely enabling change.

A further policy driver relates to the globalisation of learning: the delivery of UK higher education to worldwide markets, for both commercial and educational purposes. The formation of UK eUniversities[1] is a result of this driver. UK eUniversities is a company licensed by UK universities to deliver their higher education courses on-line. The commercial nature of the enterprise is reflected is the mission statement on the UKeU website: 'Backed by the UK Government we will deliver quality on-line UK higher education to students worldwide and improve access for students in the UK.' It is as yet too early to say what impact this development will have on the HE system as a whole.

7.3 What is the Take-up of ICT in UK Higher Education?

Any snapshot of the state of play of technology use will be incomplete, and is likely to be out-of-date in significant respects by the time a paper publication is in print. We can point to general figures indicating the state of the ICT infrastructure (such as the speed of the University networks or numbers of PCs accessible to students and to staff: the provision in the UK can probably be regarded as reasonably strong in a European context); we can indicate statistics relating to the general usage, or at least availability, of ICT tools, materials or applications (such as the 33% of departments using TLTP materials from Haywood *et al.*'s 1998 survey). However, beyond giving a general indication, such figures may be misleading in reflecting actual usage. Instead I will illustrate a few points making use of data from the discipline with which I am most familiar, psychology: there is no reason to suppose the discipline is atypical of others.

One reason general figures can be misleading is that apparently widespread usage can be patchy when inspected in more detail. An example of this comes from a survey of web usage for teaching in UK psychology departments conducted with my colleagues in 2001. Figure 7.2 shows the reported usage of the Web in a sample of 46 departments: web usage was categorised (by respondents) into four categories (plus an 'other' category where the respondent could not categorise the use) – see Figure 7.2 for details.

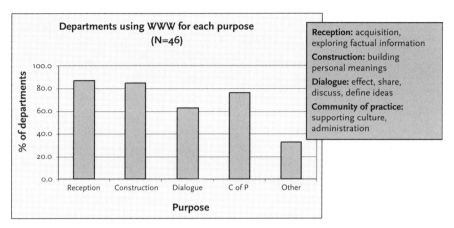

Figure 7.2 Departmental usage of web for different educational purposes

The take-up looks impressive, with the great majority of departments using the Web for each of the four purposes. However, when we asked how many courses (none, few, about half, most, all) made us of the web for each purpose, the picture changed. Figure 7.3 shows the results.

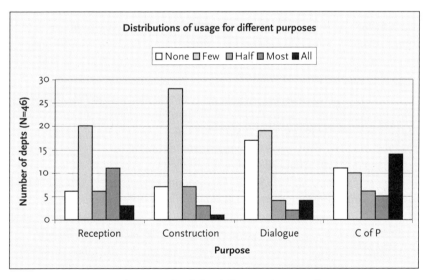

Figure 7.3 Departmental usage of web for different educational purposes

Now we see that the impressive departmental data was based on only a few courses from each department: for three of the four categories of use, 'few courses' was the most frequent response. The fourth category, 'community of

115

practice', show an interesting bimodal distribution, with the two most frequent responses being 'no courses' and 'all courses'. This reflects the fact that the category included provision of course documentation on the departmental website, and departments making information available in this way tended to do so for all courses but not for just a subset. In such cases, there are certainly gains in convenience for students (for example, the provision of a single point of access for course materials) and possibly in productivity for the department, but coverage across all courses may require widespread usage of technology by staff and their active cooperation. The more general point is that successful implementation of some applications can move smoothly from individual or a small group of enthusiasts (early adopters), whilst other applications will require a planned shift in mode of working for a large group, perhaps a whole department, cohort of students or even institution.

A related survey (reported in Hammond and Trapp, 2001) collected case studies of web usage in psychology departments. While there is not space to give details here, these case studies illustrated that much usage involves 'blended' learning, with a mix of traditional and ICT-based approaches interweaving to make up the student experience. Furthermore, at least in applications of the internet, changes in use tend to be evolutionary rather than revolutionary: there is an accretion of web usage, for access to materials, for communication purposes, and so on, rather than a wholesale substitution of traditional methods.

A final point concerning the take-up of ICT concerns discipline differences. Institutional policies can result in an implementation strategy for a technology which takes insufficient note of the needs of particular subject disciplines. For instance, use of a virtual learning environment, or of a computer-based assessment system, may be imposed across the University in a uniform fashion for reasons of efficiency. Hammond and Bennett (2002) report findings from a project (ASTER) which explored how different disciplines made use of ICT to support group-based learning. The project collected 34 case studies relating to teaching in areas of the Humanities, Physical Science and Psychology. A range of technologies was used across the case studies, as illustrated by Figure 7.4.

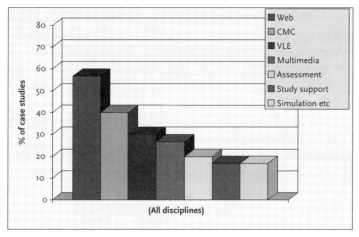

Figure 7.4 Frequencies of technology usage in small-group teaching case studies. Abbreviations: CMC – Computer-mediated communication; VLE – Virtual learning environment

When we split the same data into each discipline area, striking differences emerge (see Figure 7.5). Tools for study support and simulation are the most frequent uses in Physical Sciences; these do not feature at all in the Humanities or Psychology, for whom use of the Web and computer-mediated communication are the most common (with VLE use making a strong showing in Psychology). Further analyses showed that even when the same tools were used in different disciplines, they were often used in different ways.

117

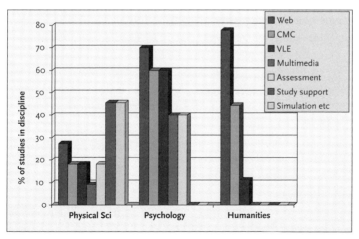

Figure 7.5 Frequencies of technology usage in small-group teaching case studies by discipline

Whilst the details of this study may be subject to various interpretations, the main point is that the same tools will be used to different extents and for different purposes in different disciplines: this is only to be expected in view of the considerable discipline differences in learning goals, learning activities and learning outcomes for students, quite apart from the distinct traditions in teaching methods across disciplines. When considering the potential of ICT, or the extent and nature of take-up, it may not always make a lot of sense to do so without considering how disciplines differ in need, existing practice and capacity for change.

7.4 Barriers, Enablers and Support Mechanisms

Ten years ago, Hammond *et al.* (1992) reported the results of three surveys exploring the perceived blocks to ICT use in higher education in the UK. It is instructive to compare their findings with a recent authoritative view on barriers to the adoption of ICT in learning and teaching in higher education (Smith, 2002). Table 7.2 summaries the top five barriers in each case. Whilst the reasons are expressed differently, there is a striking similarity across the decade: time, money and motivation are amongst the most important factors. A further interview survey reported by Hammond *et al.* in the same paper reinforces that motivation and the research/teaching balance was an issue ten years ago as it is now: 'lecturers considered making radical changes to their courses a low priority and expensive in terms of time and reward' (p. 159).

Table 7.2 Five top barriers to ICT adoption in higher education reported by Hammond *et al.* (1992; survey 1) and Smith (2002)

Hammond *et al.* (1992)	Smith (2002)
Lack of staff time	Less time because of more administration
Financial constraints	More students to teach because of falling unit of resource
Lack of staff training	Maintaining balance between research and teaching
Lack of support staff	Staff pay falling behind comparators leading to disillusionment
Lack of information on software	Staff shortages in key areas

Smith identifies a number of additional factors, including staff shortages, larger class sizes, students with varied backgrounds, staff fear of technology (still!), reward structures discouraging teaching developments and lack of institutional strategy for ICT in learning and teaching.

Turning to enablers for the adoption of ICT in higher education, there are today a host of technological enablers which were not around a decade ago. Smith, in the same paper, focuses on these: the Web, improved bandwith (enabling home and flexible working by staff and students), the convergence of leisure systems and computing systems, development of e-commerce applications, reducing cost of PC ownership and the increasing student ownership, the pervasiveness of mobile phones, the convergence of mobile phones and personal digital assistants (PDAS), and wireless technologies.

These technological enablers will be common across Europe. Within the UK higher education system, there is however a set of 'human enablers' in terms of a variety of support agencies who have the role of assisting institutions and individuals in interpreting, developing and carrying through the various strategies and policies relating to higher education practice, whether at national, regional or institutional level. The great majority of higher education institutions have staff development, educational development and computing units with the role of pushing forward institutional policy on a variety of teaching and learning matters, including the adoption in ICT. At the regional level (England, Wales, Scotland and Northern Ireland) the four separate higher education funding councils all fund initiatives aimed at promoting good practice in aspects of teaching and learning. The Teaching and Learning Technology Programme is one such example; current examples include the Fund for the Development of Teaching and Learning (FDTL) and a fund for improving higher education provision for disabled students. Grants are awarded on a competitive basis under a similar model to research funding.

A further national body, the JISC (Joint Information Systems Committee), also provides support and funding for development work[2]. The JISC promotes the innovative application and use of information systems and information technology in both further education and higher education 'by providing vision and leadership and funding the network infrastructure, ICT and information services, development projects and high quality materials for education. Its central role ensures that the uptake of new technologies and methods is cost-effective, comprehensive and well focused.' In practice, JISC support does not focus its support on individual lecturers directly but operates with other agencies and with institutions. JISC-funded projects are generally closely targeted on particular issues related to the overall JISC vision and strategy. A major element of the JISC strategy is to 'build on existing partnerships and forge new ones to contribute to a vision of a single, world-wide information environment'. The notion of the world-wide information environment is expanded in the most recent JISC annual report: 'It's one thing to have a vision of the future. It's quite another getting there. But we know the destination. The World Wide Information Environment. Information needs to be integrated with knowledge. The route is cluttered with cultural differences and technical barriers. We are using our knowledge, experience and analytical processes to help the UK further education, higher education and research sectors benefit from the creation of a global, knowledge-sharing community.' (JISC, 2001). JISC also funds a number of semi-autonomous groups and services which provide support or advice on particular topics (such as TechLearn, providing advice on new and emerging technologies for learning and teaching, and TechDis, providing advice relating to disability in higher education and further education).

Support for individual lecturers and departments is provided by a range of bodies. Principal amongst these is the Learning and Teaching Support Network (LTSN), which consists of 24 subject-based centres together with a generic centre and an Executive[3]. The LTSN aims to be the primary information and advice resource on learning and teaching matters for all academic and related staff in higher education, and its structure reflects a belief that such advice is generally best provided from a discipline-focussed standpoint. Although LTSN does not have a specific remit in relation to ICT use, the use of ICT in learning and teaching in higher education is high on the agenda of most of the LTSN subject centres. LTSN activities include brokerage (enabling the sharing of expertise), dissemination (through publications, workshops and other events) and representation to national and professional bodies.

Additional support is also provided by a number of professional bodies. The Association for Learning Technology (ALT) seeks to bring together all those with an interest in the use of learning technology in higher and further education in order to promote good practice in the use of learning technologies in education and industry, to represent the members in areas of policy and to facilitate collaboration between practitioners and policy makers. ALT provides a range of training events and conferences as well as publications and other services for its members. A more general body, the Institute for Learning and Teaching in Higher Education (ILT) is the professional body for all who teach and support learning in higher education in the UK. It exists to enhance the status of teaching, improve the experience of learning and support innovation. It also develops and maintains professional standards of practice. It intends to become the main source of professional recognition for all staff engaged in teaching and the support of learning.

7.5 Conclusions and the Future

Over the past decade, the UK higher education system has invested substantial technical and human resources in developing its support infrastructure for the use of ICT in teaching and learning. Whilst it would be hard to argue that all of this investment has been cost-effective, one would be on safer grounds in claiming that the expertise that has been built up provides the system with the capacity to develop and adapt to changing pressures in uncertain times ahead.

What are these pressures likely to be? Gazing into the crystal ball is an uncertain art, but in terms of emerging technologies one might hazard the following as particularly relevant to higher education over the next five years: the web as a major source of delivering information throughout the community; increased bandwidth and the convergence of phone and computing technologies enabling more flexible and more effective home and distance working; ubiquitous computer ownership and wireless technologies enabling more effective electronic communication and resource-sharing relating to all aspects of teaching and learning; the growth of eLearning and new education providers. Exploiting these developments effectively will be a major challenge for learning technologists. However, the technological challenges are not perhaps the most significant; there are challenges too in terms of teaching and learning practice and more broadly in terms of the educational and social role of institutions of higher education. Traditional ideas about pedagogy are already under question, both through ICT use and through other demands on the higher education system; technology may both drive changes and enable potential educational solutions to address pedagogic challenges. It will be up to educators to seek, explore and implement routes for development. The challenges at organisational and systemic levels are no less problematic: globalisation, changing markets and changing patterns of provision all raise issues concerning competition, cooperation, differentiation and the future role of traditional campus institutions. We are likely to see pressures for institutions to differentiate themselves in order to attract niche markets, and a wide range of activities and applications in relation to the use of learning technologies.

Returning to the issue raised in the first paragraph of this chapter, we have perhaps been able to shed a little light on which aspects of the higher education system may be able to adapt effectively to the rapidly changing ICT context in which we are all immersed: but only time will answer this question.

References

Brown, S., Davies, B., Franklin, T. & Smith, E. (2002). *The impact of new technologies on learning and teaching.* Presentation at ALT-C2002, 9th International Conference of the Association for Learning Technology, University of Sunderland, (9-11 September).
(http://www.techlearn.ac.uk/NewDocs/ALT%20C%202002%20Combined.ppt).

Hammond, N.V. & Bennett, C. (2002). Discipline differences in the role and use of Communication and Information Technologies to support group-based learning. *Journal of Computer Assisted Learning*, Vol. 18, pp. 55-63.

Hammond, N., Gardner, N., Heath, S., Kibby, M., Mayes, J., McAleese, R., Mullings, C. and Trapp, A. (1992). Blocks to the effective use of information technology in higher education. *Computers and Education*, Vol. 18, pp. 155-162.

Hammond, N. V. & Trapp, A. L. (2001). How can the web support the learning of Psychology? In C. R. Wolfe (ed.), *Learning and Teaching on the World Wide Web.* New York: Academic Press, pp. 153-169.

Haywood, J., Anderson, C., Day, K., Land, R., Macleod, H. & Haywood, D. (1998). *Use of TLTP materials in uk higher education – a study conducted on behalf of the Higher Education Funding Council for England between February and August 1998.* (http://www.flp.ed.ac.uk/LTRG/tltp.html).

HEFCE (1996). *Evaluation of the Teaching and Learning Echnology Programme.* HEFCE Publication reference M21/96. (summary available at http://www.hefce.ac.uk/pubs/hefce/1996/m21_96.htm).

JISC (2001). *JISC annual report 2001: A far reaching vision.* (http://www.jisc.ac.uk).

Smith, T. (2002). *Strategic factors affecting the uptake, in higher education, of new and emerging technologies for learning and teaching.* TechLearn Report. (http://www.techlearn.ac.uk/NewDocs/heDriversFinal.rtf).

Notes

1   For details, see www.ukeu.com

2   For further information on the JISC, see www.jisc.ac.uk

3   For further information on LTSN, see www.ltsn.ac.uk

8 ICT in The Netherlands: Current Experiences with ICT in Higher Education

Wim de Boer & Petra Boezerooy, Universiteit Twente, The Netherlands

8.1 Introduction

About 10 years ago, the terms ICT, Internet, on-line learning and e-learning were hardly known. This was not only the case in the Netherlands, but also in the rest of the world. Nowadays, a recent study of the OECD (2002) shows that household access to the Internet was highest in the Netherlands (69%), Denmark (65%) and Sweden (60%). Furthermore, in the same study, it has been estimated that there will be between 30 and 80 million on-line students in the world by 2025. The huge difference in estimates is due to the difficulty in defining an on-line student, since the majority of students will be studying at least partly on-line in the future. Together with an increase in the demand for on-line education, it is expected that a far more diversified and changing student profile will dominate future higher education than is the case nowadays. Already one can see more emphasis on other student groups as the so-called learner-earners (a full time student who also has a paid job) and the earner-learners (people working full time and at the same time undertaking a study).

8.1.1 Higher education and ICT in The Netherlands

The Dutch higher education system is a binary system and consists of 13 universities and about 60 institutions offering higher professional education (the universities of professional education). The latter are to a considerable extent comparable to the German Fachhochschulen or the former British polytechnics. Besides the 13 traditional research universities, there are a limited number of small 'designated institutions' that form part of the university sector: a university for business administration, four institutes for theological training and a humanistic university, as well as several international education institutes. These designated institutions also exist in the hbo-sector. Next to these two major sectors, higher education in the Netherlands is also provided through the Open University, located in Heerlen with a number of support centres around the country. The Open University offers a wide range of courses, leading to both formal university and higher vocational education degrees.

8.1.2 Governmental policy about the use of ICT in higher education

A distinctive feature of the Dutch higher education system is that it combines a centralised governmental education policy with a decentralised policy for the administration and management of higher education institutions. More and more one can see that there is a tendency towards the enlargement of the responsibilities and initiatives of higher education institutions. This implies that the higher education institutions will have much more to say in setting up priorities for higher education policies as more and more central policies are decentralised to these higher education institutions. With respect to ICT this means that there is hardly central, governmental legislation specifically related to e-learning, and that higher education institutions can decide themselves whether or not to invest in the development, implementation and use of ICT. As far as matters of intellectual property and copyrights are concerned, the Ministry of Education, Science and Culture follows the policies of the European Union (Dousma and de Zwaan, 2001). Although there is hardly any central policy with respect to ICT in higher education, there are some (semi) governmental funded initiatives that have an impact on the use of ICT in Dutch higher education institutions.

8.1.3 Surf foundation[1]

SURF Foundation is the most important Dutch cooperative organisation for higher education institutions in the field of network services and ICT. SURF was developed in the mid-1980s, with the primary goal of promoting the cooperation in the field of ICT between the Dutch higher education institutions. The aim of SURF is to provide an optimal use of ICT for Dutch higher education institutions, with an emphasis on the dissemination of knowledge. Apart from initiating innovation projects in the field of infrastructure (by SURFnet which provides the institutions with an advanced technical network on a 'not for profit' basis and SURFdiensten, delivering numerous attractive campus licenses for software and other products and services), the SURF activities can be categorised in three areas: scientific information dissemination (IWI), organisation and management (IABB), and the latest, ICT in education (SURF Education, in Dutch SURF Educatie<F>).

The SURF Education[2] programme aims to initiate, stimulate and facilitate innovative use of ICT in higher education in the Netherlands. Furthermore it stimulates cooperation in four areas: educational innovation projects, development of expertise and research programmes and the web-site EduSite (which provides an up-to-date and rich resource on ICT and innovation in Dutch higher education). In short, the major tasks of SURF Education involve the coordination of innovation projects, the dissemination of knowledge and providing a support programme to make both the projects and dissemination a success. The innovation projects comprise a diverse set of experiments and implementations of new ways of educating using ICT. Dissemination activities involve workshops, brainstorms, surveys and seminars, as well as applied research projects. The support programme places an emphasis on monitoring the lessons learned in respect of the content of the projects as well as the experiences regarding management, commitment, continuity and dissemination issues. The 3x3 matrix of SURF Education activities: 'innovation projects', 'dissemination' and 'flanking policy', are cross-linked with the perspectives of 'students', 'business' and 'internationalisation'. All this together convinced all Dutch higher education institutions to work together. A key factor is the dissemination of knowledge, to participate on both sides of the 'get some and bring some' strategy: a strategy that leads to a more generic approach, a shared knowledge base for improvement, a network of experts and peers, and results that matter to all. Since 1999, SURF Education has annually invited proposals for grants aimed at the use of ICT in the higher education sector.

125

The SURF Foundation works on the basis of a multi-year plan, renewed every four years. From 1998 to 2002, SURF Education has received governmental funds of about 17 million Euros. Innovative projects are funded with 50% matching funds from the participating higher education institutions. In addition to the 17 million Euros of governmental funding, higher education institutions have to pay a yearly contribution of 1 Euro per student. Taken together (governmental funding, higher education institutions students contribution and the 50 % matching money that institutes pay when their innovation project is selected to be subsidised) there has been a budget of about 36 million Euro for four years.

8.1.4 The Dutch digital university[3]

In the Higher Education and Research Plan 2000 (HOOP 2000) of the Dutch Ministry of Education, Science and Culture, a cutback of about 7 million Euros (15 million Guilders) in the annual governmental contribution was announced for the Dutch Open University (OUNL). This decision was based upon the decline in enrolment rates of the OUNL and on the fact that the proportion of OUNL-students that already obtained a higher education degree increased in the 1990s. The OUNL was established in 1984 aiming at providing higher education to adults that had not yet had the opportunity to attend higher education (the so-called second-chancers). In 1995 it appeared that part of this objective had been achieved, but it also pointed to the increasing student population that was not within the official target group (the second-chancers) but consisted of highly educated and highly skilled professionals, updating and expanding their knowledge base. Due to new legislation in 1997, the OUNL gained a second core function: contributing to the innovation of higher education. Although the OUNL has carried out several efforts to shape this new core-task, in the HOOP 2000 it was observed that, given the limited capacity of the OUNL and its expanding tasks, a new arrangement was needed for the organisation of distance education in the Netherlands. The need for re-organising the distance education sector was further increased due to the fact that other ('traditional') higher education institutions were also developing innovative ways delivering and improving higher education. These observations eventually resulted in two possible scenarios for the future of the OUNL:

– a merger with the University of Maastricht (located near the OUNL);
– a broad consortium in which universities, hbo-institutions and private enterprises participate.

On the basis of a scenario study of Price Waterhouse Coopers of July 2000, the Ministry decided that the consortium scenario would be the most attractive and feasible option. This resulted in the establishment of the Dutch Digital University. This consortium has two main objectives: contributing to educational innovation on the one hand and providing digital education both as part of regular higher education and for new target groups, on the other. Immediately after the choice of the consortium option, the OUNL, three universities and nine universities of professional education signed a letter of intent to participate in the consortium. The main rationale for establishing a consortium was the scale of operations. This scale was considered to be too wide for separate institutions to handle, since individual institutions had neither the resources nor the experience and expertise. Cooperation was therefore seen as a necessity to keep up with contemporary and future developments. For the participating institutions a variety of rationales can be distinguished. The DU can lead to cost reduction due to the joint development of products and through the sharing of infrastructure facilities. Furthermore, the DU provides access to new markets, to expertise that is not present within the institution and to new content. In addition, the participating institutions enjoy a 'first mover advantage' and they can make a profit by selling products to non-participants. The DU was officially established on April 6th 2001 in Utrecht. As noted above, the consortium consisted of 12 institutions in the initial phase. However, in the course of this phase, three universities of professional education withdrew their participation in the consortium. At the time the business plan was drafted (February 2001), in which the name 'Digital University' was adopted, the participants were Open University Netherlands, University of Amsterdam, Free University of Amsterdam, University of Twente, Fontys Hogescholen, Hogeschool of Amsterdam, Hogeschool of Utrecht, Ichthus Hogeschool and Saxion Hogeschool. As from 1 December 2001, the Hogeschool van Rotterdam also joined the Digital University. The first products are expected to be available at the end of 2002.

In the financial forecast of the DU, it has been estimated that the necessary contribution would amount to 9 million Euros (20 million Guilders), shared according to the status of the institution with universities being responsible for 15% (1.4 million Euros a year) and universities for professional education for 5% (half a million Euros a year). In addition, the consortium received a governmental contribution of 11.5 million Euros (25 million Guilders) for the first two years. Governmental funding for a further two years will be under consideration, depending on the results of the DU after the first two years.

127

8.1.5 ICT in Dutch higher education institutions

The technical infrastructure of the Dutch higher education institutions (supplied via SURFnet) is one of the world's fastest and most advanced networks. Speed, reliability and security of the network are the key issues. An estimated 400000 staff and students of over 200 organisations (including the Dutch universities, universities of professional education, academic hospitals, research centres and (scientific) libraries) are connected to SURFnet (www.surf.nl, 2002). Students and staff of higher education institutions can have access to SURFnet from both the office and at home. Strookman (2001) reports that 93% of staff and 81% of the students use the Internet (an increase of 15 and 9 percent compared to 1998) at their higher education institution and more and more staff and students use the Internet from their homes: almost 75% of staff and 60% of students have a home connection. Strookman (2001) also reports on the use of Internet applications: almost all users have and use e-mail. For a while, primary and secondary education institutions could make use of the services of SURFnet, but now separate new networks, like BvEnet (for the secondary vocational institutions) and Kennisnet (for primary and secondary education) have been developed for these groups.

The use of standard applications (such as Word, PowerPoint, e-mail and the Internet) but also to an increasing extent the use of electronic learning environments are part of to the standard facilities of many Dutch higher education institutions. The implementation and use of electronic learning environments is rather high (Bunjes et al, 2001). All traditional universities in the Netherlands have implemented such a system (either experiments, pilots or already institution-wide) and in 70% of the universities for professional education an electronic learning environment has been implemented. It is interesting to see that in half of these higher education institutions more than one electronic learning environment has been implemented. Most popular systems are Blackboard, WebCT and Lotus Learning Space, but also home-made systems such as TeleTOP and N@Tschool. Although many of the Dutch higher education institutions implemented (either experiments, pilots or already institution-wide) an electronic learning environment, the actual use of these systems remains as communication and organisation instead of new didactical purposes. Furthermore the availability of more extensive digital learning applications, staff development and the systematic embedding of central and faculty ICT policy can still be improved, as the results of the study described below will show.

It has become clear that, due to this kind of change in the external environment, Dutch higher education institutions have to define clear and comprehensive strategies for ICT and have to make considered choices about the markets they can and wish to serve and which type of technology they want to use. The actual influence of these external conditions, however, is determined by the way in which the internal participants perceive the changes in their environment and by their ideas about the future; there is still a gap between vision and reality. Many Dutch higher education institutions are still struggling to overcome the 'pioneer' or the '1000 flowers blooming' phase, while trying to move into a phase of more mainstream engagement. In order to be successful, indeed, the commitment of some dedicated individuals will not suffice; the institution itself must make a commitment (i.e. for support, resources and personnel) and it will have to develop a targeted implementation strategy. Finally, in order to progress both internally (involving more staff) and externally (better serving current and new students), one needs to know more about the implications of technology use.

Building on these insights, CHEPS (the Centre for Higher Education Policy Studies) and the Faculty of Educational Science and Technology of the University of Twente in the Netherlands decided in 2001 to launch an international comparative study on Models of Technology and Change in Higher Education (B. Collis & Van der Wende, 2002)[4]. The purpose of this project was to study factors that influence current models relating to technology use in higher education and which predict how institutions are likely to evolve, given their current conditions. Consequently, it explored the way in which higher education institutions perceive their changing environment in relation to their ICT strategies, i.e. which external factors are actually influencing the strategic decision-making of institutions in this area, and how they respond to these challenges. Furthermore, the study reviews how strategic responses translate into internal policies and implementation plans and what effect they are perceived to have on teaching and learning practices.

The central question for this study was: Which scenarios are emerging with respect to the use of ICT in higher education and how can future developments be predicted and strategic choices be based on that? Sub-questions were:
− what strategic choices do institutions make with respect to the use of ICT in response to these external conditions and development and how do they view their future mission, profile and market position (e.g. changing demand and target groups)?
− which external conditions and developments (changing environment, e.g. increasing competition) influence the choice of higher education institutions with respect to the use of ICT and how are these perceived and analysed by the different participants?
− what role does external collaboration play in achieving the strategic objectives (especially links with business and industry and international links)?
− which internal conditions and measures are taken in order to achieve the strategic targets (implementation strategy, role of central and de-central support units, staffing policy, etc.)?
− what are the implications of the various strategic choices or models for:

- technology use, including course management systems?
- view(s) on teaching and learning (knowledge production and dissemination) and specific pedagogical models and dimensions?
- time, workload and satisfaction of staff?

8.2 Conceptual Framework

There are many variables involved in an institution's decision to offer its educational programme in a certain way to its students. A model to study variables that influence an institution's dominant approach to educational delivery and the use of technology in this delivery will be by definition incomplete and overly simplistic. However, key variables can be identified that repeatedly have been shown to have a major impact on policy, implementation, practice, effectiveness, and eventually on an institution's general approach. We begin with what we want to predict with such a model (the outcome variables), followed by five sets of variables that can be hypothesised to have some linear relationship with each other and with the general-approach outcome variables. With respect to the outcome variables, we distinguish two main lines of change in educational delivery (Collis & Gommer, 2000; Collis & Moonen, 2001). One relates to the local vs. global issue. Should the university move toward strengthening itself as a home base for its learners, or move toward a future in which its students rarely or never come to the home campus. A second line of development relates to the programme and content to be offered. How should this be obtained, and offered to clients? As total programmes? As individual courses? As portions of courses which can be combined in different ways? Figure 8.1 gives one analysis (Collis & Gommer, 2000; Collis & Moonen, 2001).

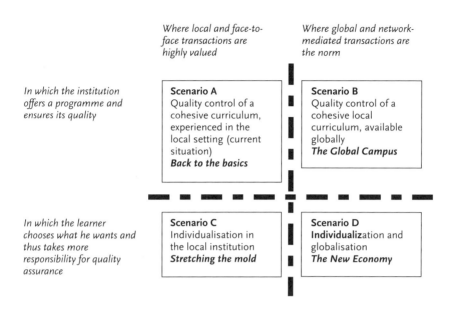

Figure 8.1 Four scenarios for educational delivery (Collis & Moonen, 2001)

Scenario A Back to Basics is the current dominant situation for many traditional post-secondary institutions. It is also the case that many universities are starting to experiment with distance participation in their established programs. This can lead to *Scenario B* The Global Campus. *Scenario C* Stretching the Mould relates to increased flexibility with or without changing the underlying pedagogical model within the institution. Many traditional universities are now moving toward some forms of Stretching the Mould, by offering more flexibility for participation within their pre-set programs. *Scenario D* The New Economy is the most radical; a systematic example of it does not yet seem to be available in most traditional universities and yet it is increasingly being seen as the way of the future.

These scenarios were used as dependent variables for the current situation of the institution but also as predictions of where the institution is headed several years in the future, such as 2005. If the scenarios are taken as the product of many different pressures and decisions within the institution, what are the main categories of such predictors?

– Environmental conditions and settings;
– Policy/response;
– Implementation;
– Practice;
– Experiences and effects.

131

We combined the aspects discussed above into a model that predicts clusters of variables that will have an influence on the current and future scenarios for ICT and educational delivery in an institution. Below the results of the study for the Netherlands[5] will be presented. (See Chapter 4 and 9 for results on Norway and Germany.)

8.2.1 *Change is slow and not radical*

This paragraph will address the way in which Dutch higher education institutions respond to their changing environment by using ICT. Before looking at the actual type of learning setting of their responses or policy, we will first analyse which external factors Dutch higher education institutions themselves perceive as being the main rationales or driving forces for these policies. In the design of the survey it was assumed that changing student demands (both changing target groups as well as changes in flexibility towards time and place for independent learning), and increasing competition and cooperation of national and international providers would be of influence on the existence of a dominant type of learning setting.

First of all respondents were asked to indicate their opinion with respect to what aspects constitute good education. Table 8.1 shows that face-to-face contact and direct communication between instructors and students are valued very highly and that in the Dutch context there is far less emphasis on the aspect of time and place independent learning (offering flexibility).

Table 8.1 Aspects contributing to good education

Aspects contributing to good education (N = 57)	Mean	(SD)
Face to face contact	4.65	(.58)
Contact with the instructor when needed by the students	4.46	(.73)
Communication among students	4.37	(.81)
Appropriate use of ICT for teaching and learning support	4.37	(.84)
Pedagogy related to group work	4.09	(.99)
Individualisation for different student characteristics	3.35	(.83)
Time and place independent learning	3.32	(1.12)

1= very little, 3=some, 5=very much

The moderate emphasis on time and independent learning (offering flexibility) is confirmed by the moderate scores (see Table 8.2) that respondents gave on the extent to which changes in student demand are currently affecting the institution's ICT policies. Here again only moderate scores for more flexibility in delivery and flexibility in locations of learning are seen. Furthermore, Table 8.2 shows that in the Dutch context there is less focus on new target groups such as international students and lifelong learners. This is confirmed by the low scores respondents gave when asked about the importance of these aspects for the mission of their institution, whereas aspects such as teaching the traditional student group (18-24 year old) and innovation in teaching and learning were valued as being of high importance to be included in the mission.

Table 8.2 Effect of changes in student demands on current and future ICT-related policy

Changing demand (N=57)	Means	(SD)
Flexibility in delivery of education	3.57	(0.98)
Flexibility in locations of learning	3.32	(1.17)
Access traditional students	3.21	(1.19)
Flexibility in pace of learning	3.06	(1.14)
International students	2.94	(1.20)
Lifelong learning	2.47	(1.07)

1 = very little, 3 = some, 5 = very much

In addition to flexibility relating to location and delivery, flexibility has also to do with the second dimension of the scenario model that concerns the extent of choice that students have in the curriculum. Table 8.3 shows that Dutch higher education institutions offer on average only moderate choice: programmes are in principle fully planned and only once students have entered the programme may they have some level of choice (a type of flexibility from within). Compared to the overall results of the study one can conclude that Dutch higher education institutions are not (yet) offering students a lot of choice related to curriculum or methods of instruction.

Table 8.3 Extent of flexibility (choice) offered, results for the Netherlands and the overall results

Extent of choice for students in the curriculum (N=693 and N=57)	The Netherlands	Overall
Fully planned programmes, some individual choices for students	51%	32%
Fully planned programmes, many choices for students	26%	39%
Fully planned programmes, little or no individual choices for students	14%	5%
Flexible programmes, students can choose from a range of combinations	7%	21%
Programmes are highly flexible, students can choose more or less own combinations	2%	3%

Besides the changing demand from students (both target groups and choices of flexibility), competition from other higher education providers (both traditional and new types) could be an external force driving the ICT policy of an institution. Respondents were asked various questions about their competitors as well as about their partnerships for cooperation in the area of ICT. The results of the study show that, as far as competition is experienced at present, it is mainly competition from other Dutch higher education institutions with hardly any from national/international business and industry. With respect to cooperation, Table 8.4 shows that Dutch higher education institutions prefer to work together with other Dutch higher education institutions, both bilaterally as well as in a consortium.

Table 8.4 Most successful forms of co-operation of higher education institutions

Forms of co-operation (N=23)	Percentage
Bilateral co-operation with national higher education institutions	39%
National consortium	31%
Co-operation with national business and industry	17%
International consortium	9%
Bilateral international co-operation with higher education institutions	4%
Co-operation with international business and industry	–

Both forms of cooperation are well-known in the Netherlands; first of all by the diverse activities employed by the SURF Foundation to get higher education institutions to work together (e.g. by annually putting out a tender for projects aimed at the innovative use of ICT in education, in which one of the criteria is that projects must be carried out cooperatively by at least two higher education institutions). Secondly, in 2001 a consortium of seven universities of professional education and three universities has been established: the Dutch

Digital University. Furthermore, new plans have been made to establish two other consortia: E-merge and Apollo in which the universities of professional education and the universities will work together on the innovative use of ICT in education.

On the basis of the above mentioned external changing environment we have asked respondents to indicate to what extent various typical learning settings occur in their institution at present and what they expect this to be in the year 2005. This question relates directly to the four choices of the scenario model (see conceptual framework).

Table 8.5: Extent to which typical learning settings occur now and in the future

Typical learning setting (N=57)	Now Mean (SD)	Future Mean (SD)
On-campus settings for course activities	4.55 (0.75)	4.23 (0.82)
Many variations in where and how students participate in courses, but campus-based settings remain the basis	3.50 (1.07)	4.14 (0.79)
Many students are attending at a distance	1.70 (0.78)	2.76 (1.03)
Students use the home institution as a base but pick and choose their courses from many locations	1.52 (0.78)	2.70 (1.12)

1=little or none, 3=some, 5=very much the case

Table 8.5 shows that, in the eyes of the Dutch respondents, on-campus is and will remain the dominant learning setting in the year 2005. It also shows that campus-based variations are moving up to become a more common type of learning setting in future (including offering more flexibility). A modest amount of change is predicted to occur related to more radical change (more distance learning students and students taking courses from other institutions), but only parallel to the on-campus mode, not replacing it. Together with the emphasis on face-to-face contact, teaching the traditional (18-24 year old) students and the demand for more flexible access and choice from traditional students for on-campus courses this confirms the 'Stretching the mould' scenario and the combination of traditional and new settings ('blended models', see below), rather than the scenarios involving ICT replacing existing practices or radically changing the traditional models and roles in the institutions (The Global Campus and the New Economy scenarios of the model). This dominant theme in the data is also supported when the way ICT is used in teaching and learning is examined more closely.

8.2.2 ICT in teaching and learning: part of a blend

Technology has become a very important aspect of education, but its importance has been overestimated through time. A lot of teaching still is rather traditional (Jorg *et al.* 2002; Collis & Van der Wende, 2002.) Although the role of technology in education has shifted from leaning towards supporting education the instructor remains very important, which is again supported by the responses on this survey. ICT use, in terms of e-mail, word-processing, PowerPoint, and the Web, has become standard as part of the teaching and learning process. But this has not radically affected the nature of this process. ICT has become part of the blend of on-campus delivery, where face-to-face contact (through sessions) and the use of ICT to support this have led to adopted ways of teaching. This trend is seen in the way that ICT use has been implemented into practice, the ways ICT is actually being used as part of a blend, and the perceived effectiveness of its contribution.

As was already indicated in Table 8.1, face to face contact still is the most important aspect for good education. The appropriate use of ICT for teaching and learning support seems also an important aspect for good education, whereas individualisation for different student characteristics and time and place independent learning are of some importance.

It seems that characteristics of the new, more flexible ways of learning do not contribute as much to good education as the more traditional ways of teaching. This is confirmed by the data as shown in Table 8.6; traditional lectures are still very common in teaching activities as well as teaching and learning in laboratories, in fieldwork or in practical exercises. Although here again, the new, more flexible ways of learning, such as studying via (non-Web) computer software and via Web-based environments are valued less than the traditional ways of teaching, it becomes clear that ICT as part of a blend, gradually stretching the traditional ways of teaching and learning, is becoming more common.

Table 8.6 The extent to which teaching practices are common

Common teaching practices (N=57)	Mean	SD
Lectures	4.67	(0.55)
Practice activities (labs, field work, practical exercises)	4.63	(0.62)
Project work, group work	4.21	(0.90)
Working via Web-based environments	3.31	(0.99)
Studying printed study materials	3.30	(1.09)

1= Not important, 3=Moderate, 5=Very important

The general level of technology infrastructure in the institutions is valued as between average and high. The available technology is used more often for organisational purposes (including course preparation) and outside classroom activities than for communication and in-classroom activities (Table 8.7).

Table 8.7 The extent to which ICT is used within institutions

Use of ICT (N=57)	Mean	SD
Course preparation or organisational purposes	4.02	(0.92)
Via a Web environment used outside of classroom activities	3.89	(1.03)
In both classroom activities AND via a Web environment used outside of classroom activities	3.47	(1.20)
For communication with and among students and instructors	3.45	(1.00)
In classroom activities	3.23	(0.85)

1=Low, 3=Moderate, 5=High

So, instructors have found useful ICT instruments and solutions to support them in their teaching practice. In particular, the use of e-mail and the use of Web resources are becoming common in educational practice, whereas other ICT forms, such as wireless solutions and conferencing tools, are used little or to a much more limited extent. Table 8.8 gives an overview of how technology influences general teaching practice.

Table 8.8 How technology influences the general teaching practice

ICT influence (N=57)	Mean	SD
E-mail systems	4.16	(1.03)
Web resources	4.00	(1.04)
Web-based course management systems	3.53	(1.40)
Planning tools, such as network-accessible agendas	2.72	(1.15)
Wireless solutions	2.16	(1.03)
Externally available courses or modules, accessible via the Web	2.16	(1.07)
Conferencing tools (video, audio, chat)	2.07	(0.88)

1=Low, 3=Moderate, 5=High

In conclusion it can be said that face-to-face interaction and direct communication between instructors and students and among students is still very important in the way instructors teach. ICT is used in a way which is complementary to this, but does not replace what traditionally has occurred in the teaching and learning process. ICT use, in terms of e-mail, PowerPoint, word-processing and Web resources, has become commonplace, but in a way that only gradually is stretching traditional on-campus practices. The lecture remains the 'core medium', the instructional form that is most highly valued. However, ICT has clearly become part of the blend, serving as a complement to already existing instructional tools. This notion of core and complementary media (Collis & Moonen, 2001) relates to the idea of blended learning, with ICT now clearly part of the blend.

8.2.3 Instructors: gradually doing more, but with no reward

What is the instructors' role in the use of ICT and how does this relate to their views on teaching and learning and on their actual workload and job satisfaction? The role of the instructors in the innovative process in an educational organisation has been extensively studied (see for example, Fullan, 1991 and Plomp, 1992). Morgan, Frost, and Pondy (1983) note that inducing change in an organisation is difficult, partly because it requires staff, both individually and collectively to change their use of language and the meanings and values they attribute to particular events. In the survey, respondents were asked to indicate how they perceive the impact of ICT use on the efficiency of teaching activities in their institution. In addition, instructors were asked to what extent ICT has increased their personal efficiency in the performance of various tasks.

Respondents indicate that the level of satisfaction among personnel in their institution with respect to their working conditions related to the use of ICT is slightly positive. The overall impression of the respondents about the impact of ICT on general working practice in their institution over the last two years is positive. Furthermore instructors feel neutral about the amount of time they need to perform for specific (ICT-related), such as using a course-management system and dealing with e-mail in their current situation.

The way instructors are supported in their change process also reflects on the use and how they perceive the impact of ICT use on the efficiency of teaching activities. Important matters here are how institutes have organised the different kinds of support, and how instructors value this. An important element in the support structure is the presence of a technical support structure, as well as the infrastructure, networks and hardware and software. The general level of technology infrastructure in the institutions is valued as highly (M=3.96, SD=1.02), where the decision makers value the level of the infrastructure higher than the instructors as well as valuing it more highly than the support staff. Another important part of the support structure is the availability of a pedagogical support structure. This level of support for instructors with respect to the use of ICT for teaching purposes in the institutions is valued highly (M= 4.20, SD = 0.95). The existence of short courses or workshops to support the use of ICT in teaching and learning are valued even higher. The results of the study show that Dutch higher education institutions score higher on these particular sorts of support than the average scores of the total number of respondents.

Staffing policy in an institution can play an important role when introducing and using ICT in education. When instructors know that using ICT counts towards promotion and tenure or that using ICT is an integral part of regular staff assessment then these will be strong incentives for them to use ICT. External quality assurance exercises can also force the use of ICT in education. Management can influence the use of ICT in education by using ICT competencies as criteria for selection and recruitment of new staff, by forcing professionalisation in ICT competencies, by financial incentives, and by declaring ICT use in education mandatory. In Table 8.9 an overview is given of the responses.

Table 8.9 The role of ICT in staffing policy (N=57)

Role of ICT use in staffing policy	Mean	SD
ICT use in education is mandatory	2.89	(1.11)
ICT use in education is part of regular external quality assurance exercises	2.64	(1.08)
Professionalisation of staff in ICT competencies is mandatory	2.53	(1.07)
ICT competencies are systematic criteria for selection and recruitment of new staff	2.49	(1.09)
ICT use in education is an integral part of regular staff assessments	2.40	(1.07)
ICT use in education counts towards promotion and tenure	2.18	(1.06)
Financial incentives to individual staff are provided for development of ICT use in education	1.92	(1.09)

1 = Not at all, 2 = a little, 3= some, 4= much, 5= very much

In general the results of the study show that ICT use plays only a modest role in institutions' staffing policy, except for ICT use in education as part of regular external quality assurance exercises. Also ICT use in education is in some cases becoming mandatory. This result shows that using ICT in education is becoming a bigger issue in staffing policy. It should also be noted that to almost all questions decision-makers are more positive than support staff and much more positive than instructors.

The instructor is 'Stretching the Mould': ICT use is part of daily practices. While there are no serious concerns about this, and a generally positive feeling about ICT's effect on personal work conditions and efficiency, there also are few or no systematic rewards to move instructors to do more than the gradual 'stretching'. Also, instructors – the ones on the front line of actual ICT use – are less impressed about it than those not on the front line. Consistently, instructors have significantly lower perceptions than the decision-makers and support staff in their institutions as to the support and incentives for ICT use.

8.3 Conclusion

The use of ICT in higher education in the Netherlands is increasing. This and other studies (Strookman, 2001; Bunjes et al., 2001; Jorg et al., 2002, Veen et al., 1999) show that although the instructors have better connection to the Internet, more tools, as well as support available, they still build upon their traditional ways of teaching. Face to face contact with students is and will stay very important. However, new ways of flexibility are explored. The Internet is used to inform students, and e-mail for flexible communication. The use of electronic learning environments is increasing.

Thus, Dutch higher education institutions do not expect any revolutionary change as a result of or related to the use of ICT. There is not really a concern about being forced to change by external forces or developments. Rather, a 'business as usual' approach is taken, in which the face-to-face contact with the traditional (18-24 year old) student groups is still very important. No real dramatic changes in mission, profile or market position are expected, especially not with respect to new target groups like international students and lifelong learners.

Nevertheless, Dutch institutions are gradually 'stretching the mould'; offering slightly more flexibility in changing their procedures, models and programs as a process of change from within. These changes, however, are gradual and usually slow and may comply with the slight changes in needs and demands as perceived by the institutions. Furthermore, it seems that the current level of Stretching the Mould is more sensitive to the level of computer use than to the particular policy of an institution.

Instructors and other participants value the face-to-face interaction and direct communication between instructors and students and among students very highly. The blend between the use of ICT and traditional ways of teaching is complementary to this and does not replace what traditionally has occurred in the teaching and learning process. The use of e-mail, PowerPoint, word-processing and Web resources, has become commonplace, in a way that only gradually is stretching traditional on-campus practices. The instructor is 'stretching the mould' where ICT use is part of his daily practices. There is a generally positive feeling about the effect of ICT on personal work conditions and efficiency, but little or no systematic rewards to move instructors to do more than the gradual 'stretching'. It is interesting to note that instructors have significantly lower perceptions than the decision-makers and support staff in their institutions as to the support and incentives for ICT use.

References

Boezerooy, P., E. Beerkens, B. Collis, J. Huisman, J. Moonen (2001). *Impact of the Internet Project: The Netherlands and Finland.* An expert study commissioned by the Higher Education Funding Council for England. Enschede: CHEPS.

Bottomley, J., Spratt, C. & Rice, M. (1999). Strategies for effecting strategic organisational change in teaching practices: Case studies at Deakin University. *Interactive Learning Environments*, Vol. 7 (2/3), pp. 227-248.

Bunjes, J., de Ronde, J., & van Wijngaarden, M., (2001). *Teleleerplatforms in Nederland. Quickscan keuze, implementatie en gebruik in het Hoger Onderwijs.* Utrecht: Surf Educatief

Collis, B. & van der Wende M.C. (eds). (2002). *Models of Technology and Change in Higher Education, An international comparative survey on the current and future use of ICT in higher education.* Enschede: CHEPS and EDTE.

Collis, B. & Gommer, E. M. (2000). Stretching the Mold or a New Economy? Scenarios for the university in 2005. *Educational Technology*, XLI(3), 5-18.

Collis, B. & Moonen, J. (2001). *Flexible learning in a digital world: Experiences and expectations.* London: Kogan Page.

Collis, B. & van der Wende, M. (Eds.). (1999). *The use of information and communication technologies in higher education: An international orientation on trends and issues.* CHEPS: Enschede.

Cordewener, B. (2001). Surfing new waves: working together on innovation. *Association for learning technology newsletter*, issue no. 34,

Dousma, T. & de Zwaan, F. (2001). E-*learning in post-secondary education in the Nether-lands.* Country paper presented at 7th OECD/Japan seminar on E-learning in Post Secondary Education, June 5-6 2001, Tokyo

Fullan, M. (1991). *The meaning of educational change.* New York: Teachers College Press.

Jorg, T., Admiraal, W., & Droste, J. (2002). *Onderwijsorientaties en het gebruik van ELO's.* Utrecht: surf.

141

Morgan, G. P., Frost, P. J., & Pondy, L. R. (1983). Organisational symbolism. L. R. Pondy (ed.), *Organizational symbolism*. Greenwich: Jai Press.

OECD (2002) OECD *Science, Technology an Industry Scoreboard 2001; Towards a Knowledge-based Economy*. Retrieved November 12, 2002, from: http://www1.oecd.org/publications/e-book/92-2001-04-1-2987.htm)

Plomp, Tj. (1992). *Ontwerpen van onderwijs en training*. Utrecht: Lemma.

Strookman, V. (2001). Internetgebruik in hoger onderwijs sterk gestegen. (http://www.surfnet.nl/publicaties/persberichten/021001.html). Utrecht: Surf.

Veen, W., van Tartwijk, J., Lam, I., Pilot, A., van Geloven, M., Moonen, J. & Peters, E. (1999). *Flexibel en open hoger onderwijs met ICT: een inventarisatie van ICT gebruik, meningen en verwachtingen*. Utrecht: Ministerie van oc&w.

Notes

1   Based on: http://www.surf.nl

2   Based on: B. Cordewener, SURF Educatief, 2001

3   Based on: Boezerooy, P., et al, Impact of the Internet Project: The Netherlands and Finland, An expert study commissioned by the Higher Education Funding Council for England, CHEPS, Enschede, November 2001

4   The project is co-funded by SURF (the support agency for technology in higher education in the Netherlands), the Bertelsmann Foundation, Germany and the Norwegian Ministry of Education. The research team consists of Prof. Dr. Marijk van der Wende (project coordinator) and Prof. Dr. Betty Collis, drs. Petra Boezerooy, drs. Wim de Boer, and Mr. Gerard Gervedink Nijhuis MSc.

5   In total, 693 respondents distributed among 7 countries, submitted the web-based questionnaires. For the Netherlands, 57 responses, covering almost 50% of Dutch higher education institutions, were submitted.

9 Perspectives on ICT in German Higher Education

Monika Lütke-Entrup, Stefanie Panke & Guy Tourlamain, Bertelsmann Foundation,Germany

9.1 Introduction

Higher education has been the focus of considerable attention in Germany in recent years. German universities and colleges face pressure from society to provide qualified personnel for a knowledge-based society. At the same time, conditions in universities have deteriorated over the last 30 years, making it more difficult for them to provide students with the education they require. While change has begun to take place, it has not occurred evenly throughout the higher education system and too often it has been reactive rather than proactive. As a result German universities are less efficient and cost effective than many of their counterparts in other countries. While there are exceptions to the rule, in general the university landscape is characterised by mass institutions, long study times, lack of direction both in the institutions and among the students, high drop-out rates and slow rates of change.

Over the last decade various institutions, both private and public, have turned their attention to the problem. Many have identified ICT as a possible solution and different forms of application have been tested in one-off programmes. Faced with a flood of individual project reports, it can be difficult to maintain an overview of ICT in universities, its potential to serve their needs and solve their problems, and the problems it brings with it. The question of ICT is inextricably bound up with the wider issue of the future of higher education as a whole. Will ICT lead to the end of the campus-based university, as was predicted at the height of the dot-com euphoria[1], or will it lead to innovations and improvements in existing institutions[2]? While there is no doubt that ICT has a role to play in the future, many of the prognoses and scenarios put forward have overestimated both the influence of free-market conditions on education and the speed of the development and spread of new technology. There are a number of structural impediments to the implementation of new media in higher education which need to be tackled before some of the more visionary ICT scenarios in higher education can be realised[3].

Following the collapse of the dot-com market, the perspectives on ICT in higher education shifted from globalisation and commercialisation back to the alma mater. This article, based on a new study undertaken by the Bertelsmann Foundation as part of an international benchmarking study coordinated by Centre for Higher Education Policy Studies (CHEPS), draws on the most recent data on ICT in German higher education. It argues that a more down-to-earth approach is necessary to reform German universities. ICT has the potential to play an important role in this reform process in five areas: teaching, research, administration, accreditation and knowledge banks (libraries, the internet, etc.)[4]. In each case, it must be approached in a realistic manner. The study shows that the promotion of ICT cannot be governed by futuristic visions of virtual classrooms, but instead by a realistic assessment of the position of German universities and the challenges and pressures they face as they strive to meet the demands of a changing society. Reform is always a slow process and introducing reform into German universities requires patience. The application of increased ICT in university curricula, institutions and administration needs to be accompanied by an understanding of the immediate requirements of German universities. ICT should be aimed at the provision of the best solutions for the problems of the present, as well as the expansion of curricula and thorough reform of structures to meet the demands of the future.

The study is the result of an on-line survey of German higher education institutions. Three different questionnaires, one for decision makers, one for instructors and one for IT support staff, were sent by e-mail to 228 institutions: 102 universities, 121 universities of applied science and 5 private universities. In total, 364 questionnaires were completed, 48 institutions took part in the survey, 21% of the institutions initially approached. Of these 30 were universities, 29% of those approached[5]; 16 universities of applied science, 13% of those approached; and 1 out of 5 private universities[6]. Given the large number of higher education institutions in Germany, compared to other European countries, the returns were considered satisfactory, especially in light of the number of surveys of higher education that have been conducted in Germany in recent years, each demanding the time of university staff[7]. Of the different target groups, the highest number of responses came from instructors, who completed 207 questionnaires. This was expected, as they comprise a higher proportion of university staff than the other two target groups. They were followed by the decision-makers with 94, and support staff with 63.

9.2 German Higher Education

In 1800 the German university appeared to have reached the end of its useful existence. In the course of the eighteenth century it had subsided, with a few notable exceptions such as Göttingen and Halle, into academic irrelevance. Leibniz, for example, complained that he felt intellectually restricted in the university environment and called for the establishment of academies and academic societies to take the place of universities. The university was, however, saved by the philosophers of the German Enlightenment. Thomasius, Wolff and Kant among others upheld its value as a place where rationalism and relativism in academic studies could thrive. This was a dramatic departure not only from the medieval, Church-dominated university, which served the needs of the Church and professed to teach an absolute truth, but also from the university in the 16th and 17th centuries, which was increasingly reliant on the patronage of the local ruler, for whom it provided trained servants.

As a result of the Enlightenment and humanism, the university became a place for free intellectual exploration. The idea that academic pursuits would reveal an absolute truth was abandoned in favour of a relativist approach, leaving the stage clear for discussion and debate.

The humanist model for the university was most famously laid down by Wilhelm von Humboldt (1767-1835), scholar, statesman and a leading figure of the Berlin Enlightenment. During a short period as the Prussian Minister of Public Instruction in 1809/10, he was responsible for the reform of the Prussian education system. He advocated the abolition of military schools and schools catering exclusively for the nobility. Instead he wanted German schools to have a diverse student population. He promoted the idea of the humanistic Gymnasium and, above all was responsible for the foundation of Berlin University, which was inaugurated in 1810. He therefore not only established a comprehensive state education system at all levels, but also a useful division of the tasks of the Gymnasium and the university, giving the latter a solid place in German society. This model for an education system still prevails to a large degree in Germany today.

The university became a state institution. The professors nonetheless retained a wide degree of autonomy. The introduction of state exams in all subjects tied the university to the state, providing the latter with graduates trained for public as well as private and academic functions. Nevertheless, by ensuring that the exams were administered by the state and that they were followed by a state training period, the university's role as an educator was distanced from entry into state service. Humboldt emphasised the importance of the pursuit of knowledge in its own right and firmly laid down the concept of the unity of teaching and research that is still fundamental in German universities today. He wrote:

> The university teacher is no longer teacher and the student no longer merely a pupil, but is rather engaged in research in which he is led and supported by the professor. University teaching not only enables an understanding of the unity of science but also its furtherance[8].

He therefore created a system in which students should learn for the sake of learning, but which did not rule out the possibility of career preparation at the same time. Professors were responsible not only for organising research but also teaching, while the students were required to learn through personal discovery and participation in the research process[9].

Some commentators have expressed doubts about the real influence of the Humboldt model on modern German universities, but it is generally recognised that it has remained the cornerstone of their self-definition[10]. While the role and structure of universities in Germany has changed and developed over the last two hundred years, Humboldt's idea of the university, and especially the emphasis on the unity of teaching and research, has survived. In 1960, the Wissenschaftsrat (Academic Council) declared that its task was not 'the development of a single system of reform for higher education in Germany.' Instead, it wanted to retain the founding principles of German higher education, which it described as: 1) the organisation of higher education as a 'community of scholars with equal rights'; 2) the unity of research and teaching and the view that the purpose of higher education is not solely the dissemination of knowledge; and 3) the linkage of specialised training and general, humanist education[11]. During the 1960s and 1970s, as the initial post-war attempt to emphasize the period between 1933 and 1945 as a break in an otherwise continuous university tradition gave way to the student movement and reform, the university landscape was changed by the establishment of new foundations and growing student numbers. Nevertheless, the changes achieved little more than the, often not very successful, remoulding of Humboldt's idea to new social and political conditions.

In the GDR, the emphasis was placed on the education of the people, and a number of new institutions emerged. All the same, socialist idealism cohabited with the Humboldt tradition, particularly in older universities such as Leipzig, Jena and Rostock. Since the reunification, the idea of the university has again been called into question, but not yet fully answered[12]. Reforms that

146

challenge the traditional vision of the university, including the reform of teaching with ICT, must therefore be approached differently in Germany than, for example, in the USA. Nevertheless, reform is urgent if German higher education is going to regain the status it once had on the world academic stage.

The roots of Humboldt's university are deep. It is upheld in the Grundgesetz (Basic Law of the Federal Republic), which protects the inner autonomy of institutions of higher education and the freedom of research and teaching[13]. Furthermore, in 1973 the Bundesverfassungsgericht (German Constitutional Court) re-endorsed the position of German universities in the state in a ruling on higher education that reaffirmed the state's responsibility to promote free academic pursuits[14]. German higher education, as a joint responsibility of the federal and individual Länder (state) governments, is also characterised by decentralisation and the principle of federalism. The national Hochschul-rahmengesetz (Higher Education Framework Law) provides a fundamental legal framework and sets out general aims and principles of higher education. It governs degree programmes, teaching and research, rules on access, membership and human resources and applies to all higher education institutions, public and private. The Länder possess discretionary powers in the implementation of the national law, which allows them considerable room for manoeuvre in this sphere.

147

Several types of institution provide higher education in Germany. In addition to the public institutions, among them 118 traditional universities, 156 universities of applied science (founded after the Second World War) and 55 art and music schools, there are currently 45 church universities and 47 private universities. The majority of institutions are members of the Hochschul-rektorenkonferenz (HRK: Conference of University Rectors), which currently accounts for 258 member institutions, at which around 98% of all students in Germany are matriculated. As the universities coordinate their activities in the HRK, which is advised by both the Länder and the federal government, the Länder coordinate their work in the Kulturministerkonferenz (KMK: Conference of Ministers of Culture). Together with the federal government, they also coordinate their higher education policies in the Bund-Länder-Kommission (BLK: Commission of the Federation and States). Both the Länder and the federal government receive advice from the Wissenschaftsrat (Academic Council). The administrative and political structures governing higher education in Germany are therefore complicated and do not allow for straightforward, centralised decision-making (see Figure 9.1).

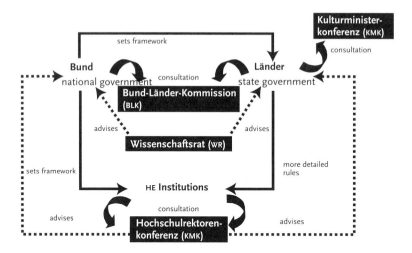

Figure 9.1 Consultative organisations coordinate German higher education
(Source: BIG)

Funding for higher education is shared by the federal and Länder governments, with the Länder providing 90% of the money[15]. The remaining 10% comes from the federal government. In recent years, funding for universities has increased slightly; per-capita, however, it has sunk. With expenditure on education as a whole at 5.5% of GDP, Germany lies below the average of 5.7% of OECD member states. German expenditure on higher education amounts to 1% of GDP, while the mean of the upper half of OECD member states is 1.3%. To enter the top 50% therefore, Germany would need to invest an additional 5-6 billion euro annually in higher education[16].

9.3 German Higher Education – Old Problems, New Demands

The problems faced by German higher education are embodied in the character of the mass university. The expansion of higher education in Germany is probably the most important change in the German education system for decades. On the other hand, it has brought with it considerable challenges for institutions of higher education, both old and new. While there was significant institutional expansion in the 1960s and 1970s, since the 1970s the number of students has risen dramatically, and many universities are filled beyond their capacity. Nevertheless only 28% of young adults in

Germany (as a percentage of the population aged between 19 and 31) enrol on a degree programme, in comparison with the OECD average of 45%[17].

The principle that higher education should be available to all social classes is strongly upheld, but studies show that social mobility is limited. While some states are currently introducing fees for those students who take much longer than average to complete their degree, in Baden-Württemburg, fees are imposed after the fourteenth semester (7th year). In July 2002 the Bundestag decided that the principle of free university education should be maintained. All the same, only 8 out every 100 children from less advantaged backgrounds cross the threshold of a higher education institution, while 72 of every 100 from more advantaged backgrounds enrol on a degree programme[18]. The significant gap between the number of students from 'academic' and 'non-academic' backgrounds, i.e. those whose parents attended university and those who did not, can be explained partly by parental encouragement, and partly due to economic and financial questions. On average, parents with degrees earn more and are therefore better able to support their children while they study.

The strengths of the German university system also contribute to its weaknesses. Once enrolled on a degree programme, the student has considerable freedom to organise his or her studies according to personal interest. True to Humboldt's ideal, there is significant room for intellectual experimentation and exploration. On the other hand, the anonymity of the large institutions, the lack of supervision and guidance, as well as overstretched facilities and unsatisfactory conditions, lead to long study times and high drop-out rates. On average, German students require about 12 semesters, that is 6 years, to complete their first degree[19]. According to official statistics, the average age of students matriculated in the winter semester of 1999/2000 was 26.5[20]. One in every four students fails to complete their degree; 16% change subject; and in some subjects the shrinkage rate is as high as 60%[21]. While in 1973 there were approximately 27 students to every professor in the Federal Republic, in 1982 the figure has risen to 36 and in 1994 to 49. The same is true concerning the overall ratio of students to academic staff. In 1973 it was 15:1 and in 1994 30:1. In the Länder of the former GDR, the situation was considerably better during the rebuilding phase immediately following Germany's reunification. A degeneration is however visible after 1994. Overall, in 1998 there were 44 students for every professor and 26 for every member of academic staff. These figures may suggest an improvement on earlier West German statistics, but are in fact skewed by the

149

new Länder in the east[22]. Moreover, the problem of high student numbers looks set to intensify in the next decade. In 2001 the KMK forecast that student numbers in Germany would continue to rise until 2010; thereafter they will begin to sink again due to the changing demographic situation[23].

The challenge for the application of ICT in Germany is, therefore, to help solve the problems of overcrowding and lack of orientation, whilst maintaining and even increasing the flexibility of degree programmes and intellectual freedom. Previously, there has either been a choice between good, individual education limited to a small number of students in a single locality and at a high cost, or cheap, mass education lacking individuality. With the help of ICT, however, this compromise can be broken as new media increasingly enable intensive, individual and cost-effective education for large numbers of people. Nevertheless, this presents the current suppliers with the challenge of providing education products that combine depth with accessibility, and at the same time avoiding the danger that their qualifications will be devalued.

In addition to the problems of the mass university, the changing needs of society are redefining the role universities are required to play in educating the citizenship. As well as providing a basic education in traditional subject areas, universities will increasingly be asked to provide lifelong learning. This will be necessary to solve the growing problem of the lack of qualified personnel in the job market, which is already a problem in Germany today. Even if the number of graduates in the natural sciences is over the OECD average, the need for qualified personnel is by no means being met. In spite of weak economic conditions, 39% of companies in industry, construction and trade are unable to fill all their open positions. Already 12% of companies are unable to expand their production or services as a result of a lack of skilled labour; some are even forced to make reductions[24]. The shortage of further education and adequate continuous training is therefore increasingly hindering growth.

The labour market requires continuous training and lifelong learning; lifelong learning in turn demands new flexible and innovative methods of teaching and training that take questions of quality and availability into consideration. It presents the institutions of higher and further education with both opportunities for new sources of income and the challenge of structural reform, calling for a new degree of flexibility that will allow people to study throughout their working lives. The results of the study show, however, that while university staff recognise the theoretical importance of lifelong learning, they consider it neither a central goal for their institutions, nor a motive for the application of ICT. Around 30% of respondents (n =343) declared that they have hardly noticed the demand for lifelong learning. Whether this reserve in the universities with regard to lifelong learning is the result of a real lack of demand from students, or whether it is motivated by a reluctance to stretch the already overstretched teaching resources of universities any further, cannot be determined from the results.

Demography and the gap between supply and demand in education and training, play an ever-increasing role influencing the job market. Lifelong learning is therefore a necessity in a time of constant social change and renewal. Traditional institutions of higher education are currently unable to provide lifelong learning of sufficient quantity or quality. New media are suitable for the provision of lifelong learning, providing both breadth and depth. It is, however, important to note that universities do not appear to see the provision of lifelong learning as an immediate goal and therefore currently are unconcerned with ICT as an effective method for its provision.

9.4 ICT and Strategic Choices

In the 1990s a system of ranking, now operated by the Centrum für Hochschulentwicklung (CHE), was introduced to German higher education to consciously promote competition between national universities with the goal of improving the quality of the education they provide. A system of formal national competition has therefore been initiated for the first time. The results of the study reflect this change. 55% of respondents (n=326) to the survey stated that competition with other German institutions had increased in the last five years. Motives for the application of ICT therefore also include the enhancement of institutional reputation and status. Nevertheless, studies show that ranking results still do not play an important role for students in deciding where to study[25].

Contrary to the expectation that the globalisation of the education market would also force universities to compete internationally, and with non-traditional providers of education, the results of the survey show they are not experiencing substantially increased international competition. Neither do they perceive heavy competition from the economic world. This is, however, likely to change. Since 1998, German industry has seen a rise in the establishment of company universities. The way was set by Lufthansa and Daimler-Chrysler, followed a short time later by Bertelsmann and the pharmaceutical company Merck. These corporate institutions provide students with a qualification specifically tailored to the job market. The Merck university offers programmes leading to degrees in engineering, economics and economic engineering. In addition students can gain an MBA, as they can by General Motors and Lufthansa[26].

An examination of the competition between German universities should also take into account simultaneously emerging cooperative initiatives among higher education institutions. The development of consortia among universities in recent years suggests that universities are aware that in order to provide a competitive service they will increasingly need to develop broader and more accessible programmes. One way to do this is to pool resources with other universities. While such structures challenge the autonomy of individual institutions to offer courses according to their own institutional design, the institutions gain from the development of information networks designed to offer students greater sources of information and increased communication possibilities. One example is Winfoline (Wirtschaftsinformatik On-line), an inter-university on-line course for business informatics, launched with the support of the Bertelsmann Foundation and the Heinz Nixdorf Foundation in 1997. Under www.winfoline.de, the chairs of business informatics at the universities of Saarbrücken, Göttingen, Kassel and Leipzig have been cooperating on the construction of an on-line learning world since 1997. At the present time, WINFOline offers 8 teaching products, i.e. on-line lectures, which cover different areas of business informatics. The plan is to broaden the curriculum further and make it available to other universities and higher education institutions. Nevertheless, the survey again showed that, on the whole, such cooperation is limited to the inter-university level, also with regard to the deployment of ICT. Cooperation with industry and the business world is still generally lacking, and would challenge the Humboldt ideal of state university education separated from business and economic gain.

Cooperation with higher education institutions on a European and international level is nonetheless increasingly considered important in German universities and a few institutions are exploring the possibilities of cooperation via the internet. For example, the International Master of Business Informatics programme offered by the Europa-Universität Viadrina Frankfurt an der Oder is aimed at professionals and students worldwide. Every stage of the programme, from admission requirements and matriculation to accreditation and course materials, is offered on-line, thereby eliminating all need to be present on the university campus. The course has been developed in cooperation with the Virtual Global University, a private initiative of 17 university professors of business informatics in Germany, Austria and Switzerland. It is promoted by the German Ministry for Education and

Research through its 'New Media in Education' programme. It therefore provides a good example of a small university with only 4000 students expanding the range of its course offering through cooperation, which enables it to use experts from other institutions to augment the expertise of its own teachers.

Competition and a service provision mentality is gaining ground in German universities, but more slowly than experts expected. Breaking the mould of campus-based learning, inter-institutional cooperation and on-line learning require new standards of excellence. For some years, institutions of higher education have been involved in discussions regarding the quality of the education they provide. New evaluation methods have already been introduced in some universities to ensure the quality of teaching, and quality management has become an important issue for universities. In January 1998, the HRK launched its 'Quality Assurance' project, its goal being the improvement of quality through the sharing of experience of quality assurance in higher education. Moreover, in order to evaluate teaching and degree programmes, the HRK and HIS (Hochschul-Informations-System) have together developed several methods. One of these, EvaNet or 'Evaluationsaktivitäten an deutschen Hochschulen im Netz', documents the various evaluation techniques and projects being used, provides information on the practice of evaluating teaching and offers a platform for the exchange of experiences.

The Ministry of Education and Research in Germany is enthusiastic to promote the application of new media in universities in order to improve standards and has provided 430 million Euro for this purpose. The majority of this money, 282.7 million Euro, is devoted to the development of educational software and a further 66.5 million to the establishment of a research network; 46 million Euro has been earmarked for degrees in IT, an area in which Germany is particularly lacking trained specialists, and the final 34.8 million has been set aside to fund virtual universities[27]. This money is augmented by the states to different degrees, with Bavaria and Baden-Württemberg both providing 50 million DM (about 25 million Euro) and Mecklenburg-West Pomerania nothing. The average state subsidy for ICT in higher education is 3.3 million DM annually[28]. Finally, the institutions themselves also provide financial support for the implementation of new media[29]. The respondents to the survey spend on average 5-10% of their budget on ICT[30]. More money does not, however, automatically mean greater success, and as far as Internet offerings are concerned the increased use of ICT in teaching brings with it new problems for quality control.

9.5 Implementation of ICT Strategy

Overall German institutions of higher education judge their application of new media to be good to average. E-mail and the Internet have gained a solid place in the everyday work of professors and students alike. New media are less frequently applied in teaching, administration and certification. Lecturers commonly use ICT in research and for course preparation, with 67% often using new media in the preparation and organisation of teaching sessions. Three-quarters expressed an interest in new information and communication technologies and a surprisingly high level of confidence as far as using technology is concerned. More than half deal (very) tolerantly with technical problems. Furthermore, the technical facilities within the institutions are generally felt to be satisfactory. Most of those lecturers asked, however, expressed few reservations about using new media.

The survey suggests therefore that the problem lies not in the attitudes of the instructors, but in the support they receive in the application of ICT, which remains limited. Over one-third of instructors described the support mechanisms in their institutions as problematic or low. In particular, didactic support for instructors for the useful application of new media in teaching is lacking.

The survey shows (see Figure 9.2) that traditional face-to-face teaching still dominates university courses. The results of the survey go on to suggest that universities are gradually adopting a path towards face-to-face teaching enriched with multimedia, a type of blended learning, which will leave university teaching largely tied to the campus in the foreseeable future, but give instructors and students greater flexibility, particularly with regard to time; the implementation of blended learning strategies could also relieve the strain on university facilities.

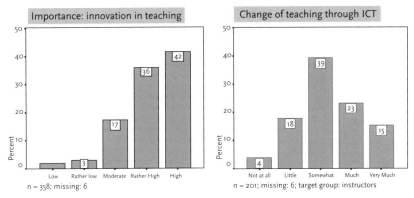

Figure 9.2 The didactic potential of ICT is not yet exploited (Source: BIG)

In Germany, teaching has always played a secondary role to the research function of universities. The study suggested, however, a reasonably balanced relationship between teaching and research, although this result should be handled with some care, being a subjective judgement by university staff themselves. While only a third of instructors regularly use ICT in teaching (n = 193), 70% of respondents stated that increased quality and innovation in teaching was an important goal in the application of ICT (n = 204). A much lower number suggested that their teaching methods were dramatically changed by the deployment of ICT.

New media offers a range of new teaching models, which could be employed in university education (see Figure 9.3). These vary from face-to-face teaching augmented with ICT, through blended learning techniques, in which face-to-face and on-line teaching are combined, to virtual universities. It is therefore potentially possible to separate teaching from the restraints of location and time, using on-line and virtual teaching, as well as telecommunication methods, providing greater flexibility in these areas.

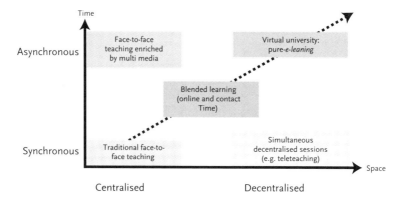

Figure 9.3 The range of ICT deployment in teaching varies (Source: BIG)

As reflected in the study, e-mail and the internet have already long been an established part of everyday university life and almost two-thirds of instructors claimed that e-mail had a big influence on their work (n = 196). 80% consider internet research very efficient (n = 195), 64% often use the internet to find teaching material (n = 202) and as many as 68% rated the influence of the internet on teaching as (very) high (n = 197). According to the estimate of technical support staff, around 40% of university staff often use bookmark functions (n = 59), whereas chat and newsgroup options are used less frequently: estimates suggest 5% use the former (n = 60) and 20% the latter (n = 61). Only 10% of instructors use course management or planning tools, or external e-learning modules (n = 193). Around three-quarters admitted that the application of computer software or web-based learning environments in teaching was unusual (n = 194). The use of ICT in the examination process and for providing students with course information is also limited, keeping students tied to the campus. The more complex the technology, the more seldom it is applied. Video, audio and chat are rarely or not used at all in teaching by over 80% of instructors (n = 60)[31].

In recent years some institutions have established schemes that make use of ICT possibilities. The VIRTUS (Virtual University Systems) project launched jointly by the Bertelsmann Foundation, the Heinz Nixdorf Foundation and the faculty of economic and social sciences at Cologne University stands, with its name, for the development of IT-based course administration. 'Virtual' means the augmentation of the traditional options offered by the university with the provision of a web-based work environment. 'University systems' is aimed at a faculty-wide concept, which gathers individual activities together and integrates them in a network applicable to the degree programmes offered in the faculty. The project is aimed at the support and improvement of teaching, service provision and communication at the university. It developed ILIAS, an open source learning management system which is currently applied in more than 200 institutions. In spite of relative success, initiatives like VIRTUS and WINFOline (see above) nevertheless remain one-off examples of blended learning concepts using on-line methods to enrich established programmes in universities. Many explorative pilot projects have not succeeded in reaching a stage of broad, university-wide implementation.

Beyond impressive pilot projects, the didactic potential of multimedia is not being exploited in German higher education. The survey showed that decision-makers identify financial resources and the lack of skilled staff as bottlenecks in further ICT development. 78% of respondents recognised the importance of innovation in teaching (n = 204) yet the majority saw little change in teaching through ICT. ICT is not seen as a means of generating increased income, but as an area in which money needs to be invested. Competence in multimedia plays a very limited role in questions of hiring, promotion and tenure in universities. Instructors, on the other hand, tend to see the lack of support structures as the main hindrance to further deployment of ICT. Over one-third of lecturers responding to the survey find the level of support for the application of ICT too low, while three-quarters point to the lack of didactic support for the application of new media in teaching. Moreover, there is a marked lack of strategic decision-making regarding ICT at the top-level. 26% of respondents said that their institution does not possess a formal strategy for the implementation of ICT or, if it does, they are unaware of it (n = 353). The majority of those that do have a strategy tend towards a combined or wholly bottom-up strategy, in which the instructors are the driving force. In many cases formal responsibility for ICT and its actual implementation do not lie in the same hands.

157

Without consistent strategies, including marketing strategies, awareness of virtual course offerings will remain low even if instructors are increasingly endorsing ICT as a possibility to enhance teaching and class organisation. Around one-third of students know that internet supported classes relevant to them exist at their university; a third are uncertain and a further third are not aware that such teaching is on offer. 30% of students know of on-line texts, booklists, exercises, solutions etc.; about a quarter actually use these on-line sources. 16% of students are offered on-line communication with their teachers; only 8% take advantage of such opportunities. On-line courses, virtual seminars, web-based training and interactive teaching are not of high importance to students; only 4% make use of them[32].

9.6 Conclusion

Change is occurring in German universities and the survey shows that universities are increasingly using ICT to enhance their performance. After several decades in which the conditions in German universities have deteriorated, they are under considerable pressure to improve. Political initiatives introducing evaluation and quality control to higher education institutions further add to this pressure. While international competition and the need for lifelong learning do not seem to have made a direct impact on the institutions, there is a growing awareness that they exist. With the introduction of ranking, national competition between institutions has already become greater.

ICT does not mean the end of the university as we know it. Flexibility and the opportunity for intellectual exploration and experimentation, the real strengths of the German system, need to be maintained and furthered by drawing on ICT to a larger degree. New demands on universities to enter the markets for e-learning and further education should take into account that the fundamental education offered by higher education will have to be the main focus. Critics often fail to see that the demands for lifelong learning and commercial course offerings expand rather than change demands on universities, which are in some areas struggling to provide the courses already on offer. The application of ICT therefore needs to concentrate on helping universities maintain their existing strengths, and provide accessible courses to as many students as possible. This also means taking into account the nature of student life in the twenty-first century, which requires increased flexibility regarding time and location.

There has been a lot of discussion about the implications of ICT for higher education. Surveys have been carried out, including the one forming the basis of this article, to find out how far and which strategies have actually been implemented. Pilot projects and one-off attempts to set up e-learning modules have been established in some institutions, in a number of cases with considerable success. The next step is to analyse the benefits of these pilot projects and to integrate best practices into mainstream university teaching. To do this two things are needed: 1) comprehensive, top-down policies, which will cooperate with bottom-up initiatives to give the latter much needed endorsement at the governmental and particularly institutional level and make the application of ICT common across the board rather than reserved to one-off initiatives; and 2) increased support mechanisms, which will, in particular, train university staff in the use and implementation of ICT in order to exploit its didactic potential to the full.

References

Arnold, P. (2001). *Didaktik und Methodik telematischen Lehrens und Lernens. Lernräume – Lernszenarien – Lernmedien – State-of-the-Art und Handreichung*. Münster.

Ash, M.G. (ed.). (1997). *German Universities Past and Future*. Providence & Oxford: Berghahn

Bentlage, U., Glotz, P., Hamm, I.& Hummel, J. (eds.). (2002). *E-Learning – Märkte, Geschäftsmodelle, Perspektiven*. Gütersloh.

Bertelsmann Stiftung, Heinz Nixdorf Stiftung (eds.). (2001). *EVALIS. Evaluation interaktiven Studierens. Studierverhalten in Präsenzveranstaltungen und mit On-line-Bildungsangeboten*. Gütersloh.

Bertelsmann Stiftung, Heinz Nixdorf Stiftung (eds.). (2000). *Studium On-line. Hochschulentwicklung durch neue Medien*. Gütersloh.

Bertelsmann Stiftung, Heinz Nixdorf Stiftung (eds.). (2001). *WINFOline. Wirtschaftsinformatik On-line, Jahresbericht 1999/2000, Erfahrungen*. Gütersoh.

Bundesministerium für Bildung und Forschung (BMBF). (2001). *Gutachten zur Bildung in Deutschland*. Berlin.

Bildungsministerium für Bildung und Forschung (BMBF). (2001). *Zahlenbarometer*. Berlin.

Brake, C. (2000). *Politikfeld Multimedia. Multimediale Lehre im Netz der Restriktionen*. Münster.

Deutsche Industrie- und Handelskammer (DIHK). (2002). *Unternehmungsbefragung*.

From Bologna to Prague – Reform of Study Programmes and Structures in Germany. Conference organised by the Association of Universities and Other Higher Education Institutions in Germany, Berlin, 5th -6th October 2000. Bonn: 2001.

Grob, H.L. (ed.). (2000) *CHL: Computergestützte Hochschullehre. Dokumentation zum CHL-Tag 2000; Alma Mater Multimedialis*. Münster.

Hamm, I.& Müller-Böling, D. (eds.). (1997). *Hochschulentwicklung durch neue Medien – Erfahrungen – Projekte – Perspektiven*. Gütersloh.

Hochschul-Informations.System (HIS). (2001). *Ergebnisspiegel*. Hanover.

Hochschul-Informations-System (HIS). (2002). *Neue Medien im Hochschulebereich. Eine Situationsskizze zur Lage in den Bundesländern*. Hanover.

159

Hochschul-Informations-System (HIS). (2002). *Studienbruchstudie*. Hanover.

Issing, L.J.& Stärk, G. (eds.). (2002). *Studieren mit Multimedia und Internet. Ende der traditionellen Hochschule oder Innovationsschub?*. Münster.

Leidhold, W., Kunkel, M. (forthcoming). *Abschlußbericht VIRTUS*. Gütersloh.

OECD. (2001). *Education at a Glance*.

Philipp, C. (2001). *Auf dem Wege zum europäischen Bildungsmarkt. Supranationale Hochschulpolitik oder Wettbewerb der Hochschulsysteme?* Lohmar, Köln.

Schramm, J. (2002). *Universitätsreform zwischen Liberalisierung und staatlichem Dirigismus. Ein Beitrag zur Theorie der Hochschulpolitik*. Frankfurt am Main.

Schulmeister, R. (2001). *Virtuelle Universität Virtuelles Lernen*. Munich.

Das Studentenwerk. (1998). *Sozialerhebung des Studentenwerks*.

Stifterverband für die deutsche Wissenschaft. (2001). *Campus On-line. Hochschulen, neue Medien und der globale Bildungsmarkt*. Essen.

Wagner, E.& Kindt, M. (eds.). (2001). *Virtueller Campus. Szenarien – Strategien – Studium*. Münster.

Notes

1   Bertelsmann Stiftung, Heinz Nixdorf Stiftung, *Studium On-line. Hochschulentwicklung durch neue Medien*, (Verlag Bertelsmann Stiftung, Gütersloh, 2000).

2   This question is addressed in Ludwig J. Issing and Gerhard Stärk (eds.), *Studieren mit Multimedia und Internet. Ende der traditionellen Hochschule oder Innovationsschub?*, (Waxmann, Münster, 2002).

3   Rolf Schulmeister, *Virtuelle Universität, Virtuelles lernen*, (Oldenburg Verlag, Munich & Vienna, 2001), p.34.

4   See Claudia Bremer, *Virtuelle Hochschule – Quo vadis?* (Unpublished manuscript for an internal publication of the gew, 1998).

5   One of the 30 was the Fernuniversität Hagen, Germany's oldest distance learning university, founded in the 1970s.

6   The rates of return reflect the number of responding institutions, not the total of members in each target group.

7   The authors are aware that statistically this rate of return cannot be upheld as representative for each target group given that it does not quite meet the necessary 30%. Moreover, given that universities of applied science outnumber universities in Germany, the results are potentially biased towards the latter. Finally, it should also be recognised that a survey conducted using the internet is more likely to attract those respondents who already make use of ICT than those for whom it plays no role in their everyday teaching. Nevertheless, we believe our results reveal some trends in German higher education and provide a useful basis for some general conclusions and suggestions for further areas of development.

8   Humboldt Universität zu Berlin: www2.rz.hu-berlin.de/ethno.english/departm/humboldt.htm.

9   Thomas Ellwein, *Die deutsche Universität vom Mittelalter bis zur Gegenwart*, (Frankfurt a/M, 1992), pp. 109-137.

10 For an interesting review of the debate see Mitchell G. Ash (ed.), *German Universities Past and Future. Crisis or Renewal?*, (Providence, Oxford, 1997).

11 Rüdiger vom Bruch, 'A Slow Farewell to Humboldt? Stages in the History of German Universities, 1810-1945' in Mitchell G. Ash (ed.), *German Universities Past and Future. Crisis or Renewal?*, (Providence, Oxford, 1997), p.5

12 Ellwein, *Die deutsche Universität*, pp.239 –263.

13 Grundgesetz für die Bundesrepublik Deutschland, article 5, paragraph 3 (2001 version:p.15)

14 Decision of the Bundes-Verfassungs-Gericht, Volume 35, p. 79 (116), from 29.05.1973.

15 BLK, Statistisches Bundesamt, bmbf – oral enquiry 2002.

16 OECD, Education at a Glance, 2001.

17 OECD, Education at a Glance, 2002

18 Sozialerhebung des Studentenwerks, 1998

19 Wissenschaftsrat 2002.

20 Statistisches Bundesamt, 2002

21 HIS Studeinabbruchstudie, 2002

22 Gutachten zur Bildung in Deutschland, bmbf 2001.

23 BMBF, Zahlenbarometer, 2000/2001

24 DIHK, Unternehmensbefragung, 2002

25 HIS Ergebnisspiegel 2002: Survey conducted in the winter semester 2000/2001.

26 See: www.karriere.daimlerchrysler.com; www.come2merck.de; www.za.bertelsmann.de.

27 bmbf, 2001.

28 his, 2002

29 Bertelsmann Foundation/CHEPS, International Benchmarking Study, 2002.

30 Median, n = 92

31 Estimation of the technical personnel (n = 60).

32 Sozialerhebung des Studentenwerks, Sonderauswertung 'Computernutzung und Neue Medien im Studium', 2002.

10 Promoting Innovation through ICT in Higher Education: Case Study Flanders

Martin Valcke, Ghent University, Belgium

10.1 Early History

Early history of the integrated use of Information and Communications Technologies in Flemish Higher Education started early in the 1970s with the introduction of the PLATO computer-system in a number of universities. The PLATO approach (Programmed Logic Automated Teaching Operations) was conceptualised at the University of Illinois and was based on the use of special terminals. The initial satellite link to a mainframe computer in Arden Hills (Minnesota, USA), was later replaced by a mainframe facility in Brussels (Belgium).

An important feature of this very first inter-university project was that it was strongly promoted as an attempt to optimise the first year instructional approaches in universities. Major applications were adapted in the field of mathematics and sciences. Although promising, the PLATO-based approach did not continue to influence higher education. This might be related to financial constraints, but also to the very behaviourist approaches reflected in the computer assisted learning programmes. Further developments reflected only in a very limited way developments in the UK, the USA and the Netherlands (Rushby, 1981; Lewis & Tagg, 1980).

We may state that this early history was clearly not related to an overall, institutional, educational policy. It was heavily influenced by research perspectives and very dependent on extra financial resources. As a result it was not unexpected that there was hardly any follow-up of these experiences during the following years.

10.2 Local Autonomy as a Limiting Factor for Innovation and Change

Higher education in Flanders went through a number of major reforms. A key milestone was the new legislation that defined the higher education arena in Flanders (1991 and 1994). This legislation makes a distinction between universities (2 cycle education) and higher education outside universities, also called professional higher education (HOBU, 1 cycle). A critical issue in this new legislation is that the government gives the responsibility to guarantee the quality of education to the higher education institutes themselves: the so-called local autonomy. The government invites the HE institutes to set up their own system of (internal and external) quality control. As a consequence, the government no longer intervenes in management of the universities and higher education institutes. The discussion about educational innovations and the use of ICT is solely a responsibility of the individual higher education institutes. This implies that, in the context of the theme ICT and innovation of higher education, the government cannot easily claim a responsibility. Moreover, due to the budgetting system (lump sum financing), the central administration has hardly any money available to set up or promote special initiatives.

The new legislation did consolidate the binary structure in higher education. This is also reflected in the way higher education institutes collaborate and play an advisory role for the Minister of Education and Training. Universities collaborate under the umbrella of the VLIR (Vlaamse Interuniversitiare Raad, www.vlir.be) and the higher professional education institutes under the umbrella of the VLHORA (Vlaamse Hogescholen Raad, www.vlhora.be).

In relation to the innovation of higher education and the use of ICT, the VLIR plays hardly any role. Only in the context of the quality assurance system, and closely related to the evaluation of specific academic programmes, do the visitation reports mention topics that question the innovation potential of ICT for specific institutes and/or courses. There is also no inter-university collaboration in relation to computing facilities that could promote discussions about ICT-use and innovation in higher education.

The VLHORA is more pro-active in this context. Recent yearly conferences have focused on ICT-related issues: 1999: Teleteaching – fostering learning ; 2001: E-learning – Didactic approaches and implementation issues; 2002: ICT and collaborative learning. Next, VLHORA acts as an intermediate for the acquisition of specific educational software and has negotiated low rates for software licenses.

10.3 Research and Development as a Catalyst for Change

Innovation processes and the educational use of ICT at higher education level are fostered by regional or national research and development organisations in a number of European countries; e.g. the Joint Information Systems Committee (JISC – http://www.jisc.ac.uk/) or the Association for Learning Technology (ALT – http://csalt.lancs.ac.uk/alt/) in the UK, SURF Educatie<F> in the Netherlands (www.surf.nl), ...

At this moment no such organisation exists in Flanders. This implies that no generic platform is available to exchange information or experiences, to report on results, to share tools, instruments or procedures, etc. Moreover, only a minority of current scientific educational research is focussed on ICT and higher education. Most of it is situated in the research department of the author of this article.

10.4 Governmental Initiatives

As explained above, in the Flemish context, central educational administration has limited impact and resources to influence institutional higher educational policies and practices. Nevertheless, in 1997 (Decree Flemish Government July 23 1997), the Flemish Government took the decision to promote innovative ICT-related initiatives with the STIHO-programme
(Stimuleringsbeleid voor de Innovatie van het Hoger Onderwijs) http://www.ond.vlaanderen.be/innovatie/Trefpunt/stiho%20index.htm). This initiative was prolonged for 4 consecutive years with an annual budget of 1 to 1.7 million Euros. In comparison to comparable initiatives abroad this budget seems very restricted. But the initiative was a real catalyst for the start-up of a large variety of ICT-projects in higher education. The project application procedure imposed a number of specifications that mirrored an approach that was never before found in this educational setting:
- projects had to be submitted by at least two different institutes of higher education;
- projects had to be focussed on implementation in the learning and teaching process;
- evaluation should be part of the project;
- projects partners had to co-finance 50% of the budget;
- project plans had to incorporate staff training activities;
- projects had to be linked to an institutional and faculty policy about ICT and educational innovation.

At content level, the programme stimulated particular ICT-related educational innovations:
- the design and development of flexible learning environments;
- the design and development of new instructional strategies and techniques;
- the development of information databases relevant for the design of learning materials;
- the design and production of generic electronic learning materials and tests;
- the professional development of teaching staff in the use of ICT.

This first large-scale initiative facilitated the launching of about 71 projects. These projects involved 45 different partners. A unique feature was that about 10 foreign universities and 2 foreign higher professional educational institutes were involved. Next, the programme seemed to have resulted in very mixed partnerships between universities and higher professional education institutes. In about 7 out of 10 projects the latter was the case. Especially promising was also the very intense collaboration between educational institutes of very different origins (state-funded, private, church, province). The STIHO-programme did also promote collaboration with international partners. As a result 12 HE partners from 3 countries were involved in the projects: the Netherlands, Norway and the UK.

If we analyse the current state-of-the-art ICT use in higher education, it is impossible to neglect the large impact of the STIHO-programme on micro-, meso- and macro-level issues.

At micro-level, the programme introduced large numbers of teaching staff to new technologies, tools, instructional approaches and materials. The output of the projects reflects the key educational ICT-related topics found in other countries: digital portfolios, simulation programmes, collaboration environments, databases with learning materials, staff training programmes, on-line curricula, special tools, etc. Moreover, the projects have been implemented in both the alpha-, beta- and gamma-sciences, with a strikingly significant number of alpha-science departments (languages, history, teacher training, logics, marketing, women's studies, etc. (for an overview see http://www.ond.vlaanderen.be/innovatie/Database/index.htm). At meso-level, the programme forced the institutes to define, at least a minimum, policy related to ICT and education. This policy implied the explicit allocation of financial resources for ICT and educational innovation purposes. At macro-level, it introduced the central administration responsible for higher education in the field of educational innovation and ICT. Departmental staff members received training and attended educational conferences.

But the STIHO-programme, although promising, also demonstrates a number of less successful elements. We discuss them in detail.

The 71 projects demonstrate a dominance of a small number of universities in the projects. The two largest universities have been leading 39 projects and are involved in 68 others. Higher professional educational institutes are only leading of 13 projects. A number of professional higher education institutes never or hardly participated in STIHO-projects.

Although the STIHO-programme clearly tried to promote the development of generic ICT-based solutions, most projects focussed on very specific solutions to support instructional problems: simulations for physics, tools for acoustics, aids for representation of Japanese fonts, etc. Only a limited number of projects resulted in tools that could be disseminated and/or transferred to other contexts.

The STIHO-programme has never been accompanied by an external quality assurance cycle. The projects have never been evaluated. No information is available as to performance indicators that could reveal the actual number of HE staff members who have been influenced, at what level instructional activities have changed, to what extent dissemination and/or transfer of tools, experiences, materials, has been realised, etc.

The latter remark brings us to the observation that, in contrast to EC R&D programmes, the STIHO-programme has not promoted a number of 'supportive measures' projects that could have taken up information exchange and dissemination between projects, between partners and in the field of HE in general. The definition of 'dissemination' activities was clearly an obligatory section in each project proposal, but the fact that each project had to do this on its own, resulted in the adoption of inconsistent solutions. The number of dissemination websites is nearly equal to the number of projects.

An attempt was made by the STIHO-programme to set up an 'Innovation of HE-portal' in 2000, but the initiative took in the month of June and hardly sufficient resources were available. In respect of information that has been obtained from the large HE institutes, it is clear that even the internal information exchange and dissemination has not been fostered. Only about 5 institutes have, for example, organised an internal 'market'or 'conference', where STIHO-projects were presented to other departments, faculties and institutional support organisations.

The selection, evaluation and follow-up of the projects hardly ever included advice as to specific technological choices. As a result, a subset of projects has developed very comparable tools, instruments and solutions. This has led to a less efficient spending of resources. But this is even truer when we see that some projects have opted to develop tools or instruments in competition with already existing ICT-solutions on the market. Clear examples are electronic course developing tools and the electronic assessment tools that are now available as consumer-market products. The in-house tools are hardly able to compete with the technical sophistication of these commercial solutions, are not able to follow the fast changes in technical specifications, standards, and operating systems, and are also hardly able to sustain internal support for the internal clients in the institute. The STIHO-project seems to reflect an ambition to become an internal/external Application Service Provider (ASP). The question is whether this should be the responsibility of an educational institute and whether this is an efficient business-model?

The key focus of the projects was on the design and development of ICT-based innovative teaching/learning solutions, but not on the implications of this new approach to teaching and learning for organisational issues, logistics, administrative processes, human resource management, financial constraints, legal issues, etc. This resulted in a number of problems related to the scalability of the solutions developed. A typical example is the fact that assessment environments are to be linked to ERP-databases or legacy-systems (e.g. student administration, rostering programmes, calendars). In the context of a project, limited on time, scope and number of students involved, this is possible. But an institution-wide implementation is blocked by technical, procedural and organisational constraints. It now becomes now obvious that a large number of stakeholders that have responsibilities that are now affected in the context of the innovative projects should have been involved in the project much earlier.

The clear focus on the design and development of innovative teaching and learning provisions has never forced educational institutes and/or departments to develop their ICT-related educational policy. The STIHO project specifications stated that the commitment of the institute should be explicitly stated in the project proposals. The validity of these 'policy statements' remains rather restricted for a large number of institutes. But this is again largely related to the overall organisational structure of Flemish higher education institutes. With the exception of the smaller professional higher education institutes, most organisations reflect a decentralised management model. The central management, the rector (vice-chancellor), the board of directors, especially define the boundaries, facilities, support structures, procedures and regulations, that influence the policies of for example the faculties or comparable organisation units. But the educational policies are defined at de-central level. This explains why the STIHO-projects hardly stimulated the explicit definition of grounded institutional ICT-oriented educational policies. A survey in the year 2000 of higher education institutes, set up in collaboration with the central education administration of the Ministry of the Flemish Community, revealed that none of them base their activities on a clear ICT-related educational policy. Hardly any examples of 'managed change' could be found.

The financing approach of the STIHO-programme resulted in 'extra' resources being provided to departments or sections in HE institutes. This 'seed' money has clearly proven to be of value and to result in a rich variety of output. But a central problem with this financing approach is that there is a strong dependency on 'extra' resources. Only a limited number of institutes seem to be able to sustain the innovative activities after the final project deadline. Staff members lose their jobs, have to re-orient their activities, licences for specific software cannot be paid, no support can be given to users of the 'products' developed, etc. This 'sustainability' issue endangers too many projects to be neglected. Many projects have invested a large amount of time and energy in the development of learning materials. Only little attention was paid to the adoption of standards to develop and store these materials. The adoption of standards compliant with the IMS specifications was hardly an issue (www.ieee.org). This affects the scalability of these projects and especially the migration to more up-to-date tools and ICT-environments.

169

The projects build on the existing organisational and operational structure of higher education. This is not always compatible with the requirements of innovative ICT-based educational approaches. ICT-applications force institutes to consider appointing staff with new and/or focussed specialised tasks (test-construction, learning material development, discussion group mentors, educational technical support line). The scalability of projects is affected by the current limitations in human resource management approaches.

10.5 More Recent Institutional Policies

In this post-STIHO period, higher education institutes observe a recurrent need for further investment in ICT-based innovations. Two major trends can be observed.

A first trend is the continuation of a STIHO-like approach but now at internal institutional level. The educational institute organises an internal call for project proposals. With very restricted budgets, faculties start up new initiatives and/or continue earlier STIHO-projects. Part of these projects are incited by comments based on the external quality control of higher education (visitations). A disadvantage of this trend is the lack of collaboration among higher education institutions and the apparent lack of expertise exchange. This trend is especially clear within about half of the universities. Recent experiences with the evaluation and monitoring of such projects reveals that this trend inherits to a limited extent the characteristics of the STIHO-projects and does not help to prevent circumvention of the less positive side-effects.

A second trend builds on the potential of generic technical solutions presented by the e-learning platforms (e.g. Blackboard, WebCT, Docent) and assessment and evaluation environments (e.g. Question Mark). This second trend is especially observed in higher professional education institutes, although recently the universities also took decisions as to the institution-wide implementation of such ICT-based learning support systems.
The strategic, organisational and operational implications of the implementation of institution-wide e-learning systems seems finally to be driving the definition and elaboration of more explicit ICT-related innovation policies.

Key elements in these policies are financial boundaries, middle-term planning, staff, centralized support provisions (help desk), infrastructure innovations in line with instructional demands, the involvement of external expertise and collaboration with other higher education partners.
A typical example of the latter is found at the University of Leuven. This university decided to implement Blackboard on a university-wide scale. This decision has been made at university level and a number of accompanying measures have been taken:
–   a special educational support unit has been set up to support all
    projects in relation to ICT and education: ICTO (Informatie- en
    CommunicatieTechnologie in het Onderwijs
    http://www.kuleuven.ac.be/icto/site/homepage.html);

- the existing computer support centre was re-organised to become a service-organisation that supports access to the computer facilities at the university, in the university hostels and at home (Ludit – http://ludit.kuleuven.be/algemeen/);
- ICTO and Ludit are responsible for the Toledo-project (http://toledo.kuleuven.be) that supports the implementation of both Blackboard and Question Mark. Staff members find help, training, background informa-tion on instructional approaches, individualized training packages and can contact a helpdesk.

Project staff members support major procedures for teaching staff; e.g. the migration of available courses to a new academic year, subscription of students, etc. June 2002, the e-learning environment has been used for one full academic year. Nearly 700 courses are operational involving about 16000 students, covering an equal proportion of undergraduate and graduate level courses. 85% of the use is for information exchange; only a limited number of on-line courses foster communication and interaction. Exploitation of communication features is more observable in the humanities. Top user faculties are applied sciences, medicine, sciences and arts.

Acceptance by students and staff members is high. Key problems are not reported at micro-level, but rather at meso-level in relation to administrative and technical issues. The integration with LDAP (ERP-database) is not yet completely finalised, causing difficulties with the authentication of students and staff. The integration of the e-learning environment and the assessment environment is not complete.

An interesting sub-project of the Toledo-programme is the future integration of the Ariadne system that fosters the re-use of learning materials by adopting IMS-compliant storage standards (http://www.ariadne-eu.org/).

10.6 The Central Approach is 'Blended Learning'

The current state-of-the-art in Flemish higher education as to the integrated use of ICT in the educational process can be qualified with the concept 'blended learning'. This 'blended' approach reflects the adoption of a variety of media, instructional formats, organisational choices and a mixture of syn-chronous/asynchronous, dependent/independent learning and face-to-face/learn-ing at a distance. It can be argued that this 'blended' approach is but a stage in a gradual evolution towards more radical 'virtual learning organisations'. But this is countered by a number of concrete observations. For instance, in the most recent version of the new educational policy of the Ghent University, we read:

> The Ghent University is an institution that grounds its educational approach in the face-to-face contact between learners, masters and experts. The inherent quality and importance of this personal contact recognized. is (...) The Ghent University opts to present itself as a 'campus university' (Ghent University, 2002).

But the same policy paper also clearly opts for the implementation of a campus-wide e-learning system, the integration with back-office systems and the fostering of a-synchronous teaching and learning approaches.

As to the latter, the Catholic University of Leuven presents again a clear example of this blended approach. In a recent evaluation study researchers involved in the TOLEDO-project comment on of the (Buelens et al., 2002) have analysed the log-files that registered all the key strokes of users of the Blackboard environment. They distinguished between three clusters ofdidactic uses of the e-learning environment. Figure 10.1 represents the proportions of didactic usages during the first and second semester of 2001-2002. The classification in Figure 10.1 reveals that, of the 571 courses, the majority is 'document oriented', i.e. focussed on the distribution of information and resources about the course or supporting the course content (Power Point slides, agenda, rostering, news, frequently asked questions, etc.). This picture remains consistent during both semesters. Initially the researchers were rather disappointed with these results. Their assumptions about the potential of the e-learning environment to foster innovative teaching and learning approaches were not confirmed by these data. The data show that the e-learning environment did rather reinforce existing traditional educational practices. The orientation towards communication-related usage is very limited. But, the picture is completely different if we consider how the e-learning environment is used in close connection to the face-to-face sessions. The research did not study in detail how these sessions were set up. Informal discussions with users of the Blackboard environment suggest that the focus on communication and interaction has not yet been transferred to the use of the e-learning environment. Teaching staff seem to prefer to foster this interaction and communication during personal meetings with their students. As a result, the 'blending' of the e-learning and face-to-face approach does represent a fuller picture of the teaching and learning models adopted in the institutional context.

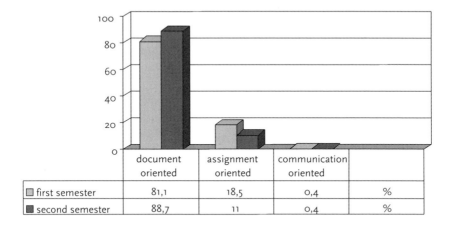

	document oriented	assignment oriented	communication oriented	
first semester	81,1	18,5	0,4	%
second semester	88,7	11	0,4	%

Figure 10.1 Analysis of key strokes in terms of didactical usage of Blackboard

Another example is projected in the future plans for the new teacher training approaches. To cope with the current shortage of teachers in primary and secondary schools, blended learning is proposed to support student teachers who already start working while they study. This study process is fostered with e-learning provisions.

10.7 The Impact of Recent Macro-level Policies

This picture of educational innovations based on ICT use is currently again affected by the gradual definition of some major changes in the structure of Flemish higher education. In view of the introduction of Bachelor-Master structure, the Minister of Education has invited both universities and professional higher education institutes to create 'associations', i.e. clusters of institutes that can guarantee flexible integration and/ or compatibility of instructional programmes. Although this new development is far from a discussion of ICT and educational innovation, there is a very clear link. The fact that a number of higher education institutes have recently chosen a specific e-learning platform and/or assessment tool, influences the emerging partnerships. The partners bring into line their technical, organisational and operational choices. A key one, affecting the organisation of instruction, is the integrated use of ICT and especially the e-learning platforms. Pricing, licensing, etc. are now taken up in the context of these much larger 'units' of higher education. The impact of this development on higher education will be very obvious in the very near future. The example of the University of Leuven has already been discussed. Examples in higher professional education institutes can be found at, for example, the Provinciale Hogeschool Limburg (http://www.phlimburg.be/) and the Karel de Grote Hogeschool Antwerpen (http://www.kdg.be/english/index.htm).

A very recent initiative of the Flemish Government in July 2002 focuses on the universities and invites these institutions to submit proposals related to the optimisation and innovation of their education (detailed information via http://www.ond.vlaanderen.be/hoger_onderwijs/OPROEPRECTOREN%20ministe rie_van_de_vlaamse_gemeen.htm). An exact translation of a particular paragraph of the tender text indicates how much thinking at the macro-level has evolved during the last two years:

> (...) The following projects can be submitted: (...) Projects related to the inno-
> vation of education and quality assurance by the redesign of curricula, the
> pedagogical-didactical approach to the learning process, the design and deve-
> lopment of new learning materials, the design and development of new eva-
> luation and assessment approaches, new types of student support, potentially
> including the development of electronic learning environments in the context
> of full-time and part-time studies.

The call for proposals obliges the institutes to submit projects as a comprehensive set of activities that are in line with an explicit educational policy.

10.8 Research and Development

In Flanders, no centralised organisation or national body is involved in the monitoring of ICT in relation to education. The recent report on the status of higher education (Departement Onderwijs, 2002) incorporates some general reflections about, for example, the STIHO-projects, but these comments are of a general nature and are not based on a systematic review of all projects. This approach results, at macro-level, in a lack of information about the actual state-of-the-art, about the availability of infrastructure in the institutes and/or at home (students and staff), no information about current levels of integrated ICT-use, the needs for professional developments, the impact (efficacy, efficiency, satisfaction) of ICT-use, etc. The information presented above is as a consequence not grounded in large-scale studies about the current state-of-the-art, but rather builds on the experiences of the author of this contribution by being heavily involved in these developments.

A small number of educational institutes have adopted an organisation structure in which a central unit monitors the issues discussed here. It is therefore difficult for higher education institutes to have a clear insight into the actual position of their instructional practice in comparison to higher education in general and/or partners in particular. As a result, only a limited number of local studies is available that focuses on departmental developments, evaluation of particular aspects of the ICT-based innovation, etc.

We have already referred to the research related to the TOLEDO-project of the Catholic University of Leuven that focused, for example, on the analysis of the log-files when using the Blackboard environment. At the Ghent University (www.rug.ac.be), the Department of Education focuses on research into the implications of ICT-based innovations on student perceptions, study skills and the congruency between characteristics of these learning environments and learning styles of students (see Schellens & Valcke, 2000). At present, especially the impact of collaborative learning environments (computer supported collaborative learning) is researched (see De Wever, Valcke & Van Winckel, in press). The results of the researches indicate that HE institutes should be careful when introducing innovations at the level of single courses. To have a larger impact on students and to influence or develop learning styles in a consistent way, the results present a plea for an innovation at curriculum level or at least at the level of a set of interrelated courses. Another conclusion drawn from this research is that not all freshmen entering higher education are 'ready' for specific innovative approaches. Care should be taken to orient these students towards study behaviour that is in line with the expectations of the ICT-based learning environment.

175

10.9 The Way Forward

Despite the fact that ICT-based innovations in Flemish higher education builds on only a very recent 'history', and the fact that comprehensive research-based data are lacking, a number of lessons learned can be formulated. We distinguish between macro-level and meso-level recommendations.

At macro-level, the experiences suggest that policy makers should foster collaboration between institutes. The size of the investments in financial and personal resources demands such collaboration. Moreover, institutes are in this way able to combine mutual expertise. Policy makers should also create incentives to set up international partnerships that focus on ICT-use in higher education. Flemish higher education institutes could leap forward if they build on the expertise already available abroad.

Next to this, 'seed money' remains crucial. The newest initiative of the Flemish government to foster innovation in university higher education is in line with this 'lesson learned'. This extra money is crucial to start up innovations in HE. But next to this, history has also shown that accompanying measures should be supported, and this to a large extent. The latter is especially true for support actions such as dissemination, expertise exchange, evaluation and quality assurance, etc.

Next, institutes should be refrain from investing high amounts of money in technological developments. Policy makers should foster conditions for the adoption of industry standards and consumer market products. Staff training should be stressed in the policy documents and be an integral part of each project at micro- and meso-level. Since higher education institutes already work together in the context of umbrella organisations, these organisations could also become a partner to promote ICT-based innovations.

The development of evaluation capacity is crucial. Policy makers should start up and promote national evaluation and monitoring activities. At the same time they should foster the development of this evaluation capacity within the higher education institutes. This could result in the definition and adoption of adapted internal and external quality assurance cycles.

At meso-level, the lessons learned request institutes to adopt a number of strategic, organisational and operational measures. Above all, there will need to be a continuous effort in developing an educational policy, including a policy focussing on ICT. This policy has to reflect an integrated approach towards a number of issues. We present the following list that could be derived from the lessons learned:
- foster collaboration within the HE institute (between faculties, departments);
- focus on initiatives that are scalable;
- extrapolate the administrative, logistics, technical consequences of ICT-projects. This implies that the projects should start as early as possible to

176

negotiate with departments that possess crucial information about students, staff data, rosters, calendars;
- set up an internal quality control cycle;
- foster collaboration with other higher education institutes; e.g. in the context of 'associations';
- invest in a significant way in central support (helpdesk, training, documentation, registration, authentication);
- discuss and determine the consequences of educational ICT-use for the human resource management: incentives for staff, staff development, specialisation, new job profiles, etc.

If we consider these lessons learned in view of the most recent initiative of the Flemish government related to the innovation of university-based higher education, we can observe that a number of the lessons are taken into account, especially the major importance attached to a central educational policy and the fact that the projects have to be submitted as an integrated institutional programme.

10.10 Conclusion

As stated earlier, the adoption and integration of ICT in Flemish HE is still fully under development. The biggest contrast with developments elsewhere in Europe and the world is the lower level of attention paid to policy-based decisions and the restricted involvement of national bodies and organisations that direct relevant developments. The case studies show how initial developments were based mainly on micro-level initiatives. But, the national STIHO-programme, although limited in time and financial size, functioned as a catalyst in a process of the development of local policies, institutional collaboration and the development of ICT-related educational expertise. The initiative has helped to put 'ICT and educational innovation' high on the agenda of institutional policy makers.

It is a pity that the STIHO-project was not monitored in detail. Nevertheless, ICT in Flemish HE seems to have taken off, but largely depending on bottom-up approaches. A promising development is related to the institution-wide implementation of e-learning platforms and this in the context of the growing 'associations' between HE institutes in view of the adoption of the European Bachelor-Master structure.

But, this critical review of the integrated use of ICT in Flemish Higher Education would not be fair if it did not end with a positive comment. Despite the fact that Flemish higher education institutes reflect a less systematic and policy-based approach, emerging practices mirrors a very high level of enthusiasm and involvement of staff members. The innovations are grounded in the personal motivation and clear professional orientation of staff members and teams. From this perspective, we can state that future ICT-related educational policies at macro- and meso-level can be grounded in sound foundations that rely on the necessary human resources.

References

Buelens, H., Roosels, W., Wils, A. & van Rentergem L. (2002). *One year E-learning at the HULeuven: an examination of log-files.* Paper presented during the Conference The new Benefits of ICT in Higher Education, September 2-4 2002. Rotterdam: Cheps, OECR, SURF.

Departement Onderwijs (2002). *Het Jaarverslag over het Hoger Onderwijs in Vlaanderen 2000-2001.* Brussel: Ministerie van de Vlaamse Gemeenschap. Retrieved from http://www.ond.vlaanderen.be/hoger_onderwijs/univ/jaarverslag%202001/start-pagina.htm on September 20 2002.

De Wever, B., Valcke, M. & van Winckel, M. (in press). The impact of 'structure' in CSCL-environments: a study with medical students. The Medical teacher.

Ghent University. (2002). *Visie op onderwijs als antwoord op de veranderende behoeften vanuit de kennis- en informatiemaatschappij.* Internal discussion paper, version September 11, 2002.

Lewis, R. & Tagg, E. (1980). *Computer assisted learning.* London: Heinemann.

Rushby, N. (1981). *Selected reading in computer-based learning.* London: Kogan Page.

Schellens, T. & Valcke, M. (2000). Re-engineering conventional university education: Implications for students' learning styles. *Distance Education,* Vol. 21(2), pp. 361-384.

Coordinating organisations

- Association for Learning Technology: alt http://csalt.lancs.ac.uk/alt/
- Ghent University: www.rug.ac.be
- ICTO: http://www.kuleuven.ac.be/icto/site/homepage.html
- Joint Information Systems Committee: jisc http://www.jisc.ac.uk/
- Karel de Grote Hogeschool Antwerpen: http://www.kdg.be/english/index.htm

- Ludit: http://ludit.kuleuven.be/algemeen/
- Provinciale Hogeschool Limburg: http://www.phlimburg.be/
- STIHO-projects: http://www.ond.vlaanderen.be/innovatie/Database/index.htm
- surf Educatie<F>: www.surf.nl
- Toledo-project: http://toledo.kuleuven.be
- VLHORA: www.vlhora.be
- VLIR: www.vlir.be

11 Expectations, Opportunities, Achievements, Realities: ICTs, eLearning and ODL at Universities in Hungary, with a Look at Central and Eastern Europe

András Szucs, Budapest University of Technology and Economics, Hungary

11.1 Introduction: The Modernisation Challenge and ICTs in Education in Central Europe

Since the beginning of political and economic changes in the early 1990s in Central and Eastern Europe, the continuous and demanding mission of development and restructuring the economy and catching up with the better developed part of Europe, also the challenge of joining the EU, with all the structural, economic and compatibility requirements, have produced an uninterrupted challenge in the countries of the region.

The impressive progress of information and communication technologies has largely influenced worldwide the development of economies and transformation of cultures, but also our daily lives in the same period of time. For the economies in transition, modernisation of the economic and social system, strengthening the competitiveness and meeting the requirements of the European Union accession in the accelerated environment had to be managed together with the information technology challenge.

Education has been, particularly in the first half of the nineties, an important element in this process. The political and economic changes in Central and Eastern Europe were met by the academic community with enthusiasm and great expectations concerning European integration. Higher education seemed well positioned to play a flagship role in the development of East-West cooperation, as the difference between the university spheres of developed countries and Eastern Europe was certainly less than between the industrial or commercial sectors. The European Union was quick to offer support, to be delivered under a multi-sector programme known as Phare. The first assistance projects within Phare started in the field of higher education cooperation (TEMPUS). These developments have sent a significant message to the higher education sector in Central and Eastern Europe: the European Communities consider that higher education could take the lead in cooperation with European Union Member States and that universities could be expected to play an outstanding role in the transformation of society and economy.

When the eEurope initiative of the EU was launched in 1999, the message and the order of magnitude of the challenge was again well understood in the candidate countries. The need for awareness concerning the governmental responsibility in initiating and supporting the development was acknowledged and a number of national actions initiated accordingly.

Studies investigating in Hungary the starting position before the planning and implementation of comprehensive actions for the development of ICTs in the second half of the 1990s (UNESCO, 2000) concluded that the main characteristics at that time were the lack of strategic awareness, the badly underfinanced status of key fields of the information society on one hand, the excellent professional background and the existing good networking infrastructure in the higher education and research field on the other. There was a missing institutional background at government level for the support of development of the information society, and this deficiency was accompanied by the related lack of economic priorities. The necessity of strategic and institutional development was acknowledged.

11.2 Development and Recent Status of Using Information Technologies in Education in Hungary

A current survey about eLearning in Hungary (Eduweb Co, 2001) states that notable eLearning solutions can mostly be found in higher education and in the corporate sector. The appraisal of the existing practice of distance education and eLearning is in the meantime rather diverse. As a result of the efforts and financial resources invested in the development of open and distance learning in Hungary since the early nineties, with a dominant contribution of EU educational programmes support (mainly TEMPUS), there are 20-25 well equipped and professionally prepared institutions. Their activities are however mosaic-like, the critical mass or the coherent functioning/networking, identification and exploitation of synergies is missing, which reduces the efficiency of efforts. Besides a few internationally acknowledged centres of excellence, initiatives and activities of questionnable quality are also present in the market. A number of companies are offering eLearning solutions and developments, at least sold under this name, but not always meeting the professional criteria of this form of education.

Most of the significant eLearning platform and learning environment softwares are available and in use on a small scale in the country. No Hungarian language version of these products has been developed yet. There exist few platforms or learning support systems developed in Hungary offering services comparable with the already known and introduced products and services. Experts assume that the number of platforms used will be reduced because the Hungarian market is in terms of size quite small and the courses developed not compatible with each other. In the field of course development, most eLearning courses developed under one of the main platforms are predominantly in English. Courses offered in Hungarian can mainly be found in the field of computer science or management (Consulting & Research for Industrial Economics Ltd, 2001).

During the past half decade, the size of population in Hungary decreased in all age groups concerned with public education. The capacity of the educational systems did not follow this trend; thus student/teacher ratios have improved, but cost efficiency has decreased. Resources have been slightly increasing (beyond the inflation rate) in all levels of education. In ICT related developments there is a considerable widening of governmental resources for education, both regarding the number of initiatives and the amounts of resources available.

The ICT market in general has been showing a fairly rapid development during the past few years in Hungary. The increase in the number of main phone lines has, following a very dynamic development in the mid-1990s, slowed down in around 2000. Mobile phone penetration has developed very intensively during the past few years, exceeding already the number of traditional phones and has resulted in a quantitatively almost saturated market. There is a somewhat unbalanced situation concerning computer penetration and Internet access: whilst in the education and public administration sectors there is a satisfactory situation, the number of PCs and Internet use at home is significantly behind the EU rates. In the autumn of 2001, 18% of the adult population, i.e. 1,480,000 people had access to the Internet, but the number of Internet subscribers was 282,000, out of which 7.7%, (22,000) came from the corporate area.

In the first half of the 1990s the IT infrastructure of higher education institutions developed considerably using as main resources EU funds (TEMPUS), World Bank loans, the 'Catching Up with the European Higher Education' national programme and the National Information Infrastructure Development Programme. In the university sector, surveys (Consulting & Research for Industrial Economics Ltd, 2001) show that there is practically full access to computers and the Internet among the teachers at the faculties. About one-fourth of PCs in the higher education sphere in 2001 had 8-10 year old configuration, applicable for simple word processing or e-mailing purposes only. Pentium I-II computers represent 60% of the machines used. For purposes of modern IT education, about 75% of the computers available are appropriate.

Concerning recent levels of investment: higher education institutions spent in 2000 about 6.5 MEUR for IT equipment and services. Most of this amount was spent on hardware (80%), for software about 1MEUR (15%) and for the development of web sites, a small proportion only. The forecasted increase in IT investment for future years is about 8%, with about double the amount for software compared to hardware and 6-7 times (!) more for website development. The maintenance and service costs related to the IT infrastructure represent about 20% of the yearly IT investment.

184

All higher education institutions maintain their own website and on average update them at least monthly. According to the surveys, 74% of the students use the Internet to supplement the information received in the face-to-face teaching process and 78% of the teachers publish courseware or assignments on the web. More than 50% of the teachers use IT-based support developed by themselves in education. Internet and internal networks are intensively in use for the purposes of institutional administration, management of international affairs, and information gathering by teachers or administrators at the universities.

Computer courses and basics of informatics are taught to students in higher education for an average of 109 hours during the whole period of education. In colleges, this figure is 133 hours, whilst in universities only 82 hours were identified.

11.3 The Policy Dimension – Legal and Institutional Environment, Governmental Programmes for ICT Development in Hungary

11.3.1 Institutional background and legislation

From the second half of the 1990s, an increasing governmental involvment has been present in the comprehensive development of ICTs, including legal regulation and economic policy aspects. This resulted in 2000 in the establishment of the 'Office for Government Commissioner in charge of Information Technology' to ensure the institutional background of development actions and the elaboration of a comprehensive development strategy. In May 2001, the National Information Society Strategy has been published (Office for Government Commissioner in charge of Information Technology, 2001). At the end of 2001, the 'e-Government' portal was opened[1].

After the parliamentary elections held in April 2002, the new government has identified a number of priorities carrying strong modernisation elements. An important message was the establishment of the Ministry of Informatics in the new governmental structure, replacing and positioning on a higher level the tasks of the earlier Office for IT Government Commissioner.

The ICT-related intensive legislative work resulted in acts in the fields of civil law, economic and commercial law, consumers' rights protection, intellectual property rights, etc. A recent list of acts relating to ICT in Hungary includes an Act on the Post (XLV/1992), an Act on the Telecommunication(LXXII/1992), an Act on Frequency Management (LXII/1993), an Act on Radio and Television (I/1996), an Act on Communications (XL/2001), an Act on the electronic signature (XXXV/2001), an Act on the electronic commercial services and on certain issues concerning other services related to the Information Society (CVIII/2001), and the Act on Adult Education (Act XI/2001). These legislations, supplemented with other governmental and ministerial decrees, form a basically coherent and supportive environment for the development of the information society.

11.3.2 Comprehensive national programmes and their educational dimension in hungary at the turn of the millennium

The initiatives listed below show well the increased awareness of the Government in the late 1990 concerning the need for broad actions at national level in support of development of the information society in Hungary. These activities carried the message of modernisation using ICTs and with the widening of resources helped to reduce the existing digital lag of the country. The evaluation of achievements of these schemes had not been done when this article was being written. The new government which has been in power since the middle of 2002 has in the meantime identified in this field a number of new priorities and implementation modalities, launched new institutions,most importantly, the Ministry of Informatics and programmes. These years will in any case be noted as a period of very significant and intensive development in the field of ICT development in terms of widening the access to tools and facilities, establishing the important national frameworks and professional conditions which help the appropriate and good quality use of information technologies, and moving towards the reach of the critical mass concerning their application in all relevant spheres of society.

11.3.3 National strategy for information society

A predecessor of an elaborate national information society strategy was the 'Theses about the Information Society' paper (Prime Minister's Office, 2000). This policy document highlighted the importance of knowledge and the emergence of the knowledge capital as the determining factor in the development of the society, the need for organic incorporation of the new technologies in the learning process, the significance of lifelong learning and the need for teaching informatics at all levels of education.

The National Strategy for Information Society (Office for Government Commissioner in charge of Information Technology, 2001) declared that Hungary is behind the leading countries of the information age and, even if the growth of the GDP and the development of the infocommunication market is higher than the European average, the lag to be eliminated in the fields of technology standards, intelligent applications, the market environment, the readiness for the information society and the indicators of Internet access and further use, in the field of eBusiness solutions is still considerable. Catching up with the developed countries needed active governmental action, along a consistent policy. As the Strategy states, by the end of the 1990s, Hungary has reached the position to consider the ICT challenge not just as a threat but as a chance for break-out and improvement of the international position of the country.

The strategy document has a structure of main chapters of sub-programmes for infrastructure development, economic policy, culture, education, social policy, e-government and local government. The new attitude applied in the national strategy can be well characterised by the approach that not only public institutions but also corporations and non-profit institutions should be the beneficiaries who are expected to increase further the outcomes of the financial support with multiple effects and by ensuring the sustainability of actions.

The *educational sub-programme* of the National Strategy identifies as key fields the following: research and development in educational methodologies for the efficient use of new technologies, particularly in open and distance learning, course and content development in different levels of institutions in the field of ICT-supported education, in harmony with the methodological developments, development of courseware improving the digital literacy and supporting the IT equipment provision for schools.

187

11.3.4 Széchenyi plan

The Széchenyi Plan, named from the 19th century legendary Hungarian statesman István Széchenyi, has been announced as the middle-term (2001 – 2006) economic development plan of Hungary. ICT and education/research in the Széchenyi Plan are mentioned in two main strands: the Programme for Research, Development and Innovation in the information society and information economy development. For the support of these programmes, the state budget for the years 2001 and 2002 ensured 126 MEUR and 256 MEUR, respectively[2].

Main objectives of the information society and information economy development program of the Plan include: the promotion of the establishment of electronic public administration to create the legal and regulatory background of the information society and information economy, improvement of availability of up-to-date ICT equipment, to increase Internet access; and the support of the Hungarian language content provision and information services.

11.3.5 Apertus public foundation for open vocational training and distance education

The Government founded in 1997 the Public Foundation for Open Vocational Training which continued from 2001 as the Apertus Public Foundation. The Hungarian National Council for Distance Education works as advisory professional body for the preparation of decisions and development of the strategy of the Foundation. These foundations have been able to rely on the financial support of the 'vocational training contribution' paid in the state budget by employers, which ensured a considerably higher funding than the earlier rather scarce governmental support for open and distance learning through the Ministry of Education. The Apertus Public Foundation has disbursed in 2001 in the frame of application procedures 4.4 MEUR, with explicit preference for the development of on-line e-Learning solutions, courses and programmes but also for the development of national frameworks of methodology, evaluation and quality assurance[3]. A number of higher education institutions could be found among the beneficiaries.

The major strategic goals of the Apertus Public Foundation as mentioned in the Deed of Foundation are the following: to extend the range of programmes and training possibilities for adult training, to develop the distance education system in vocational training by using and developing IT tools, to improve the labour-market position of the employees by using the methods of distance education, to develop and operate a quality assurance system for open professional training, to support the development of international relations of the professional area, to create a national consultancy network and examination system, to develop the IT background for the delivery of programme packages, and to support Hungarian communities living over the border for the launch of training programmes in the Hungarian language. Emphasis during the implementation of the above goals is on the introduction and adoption of the progressive European practice of open training and distance education, modern technologies to support training programmes and by supporting applied research in training.

11.3.6 IKTA – information and communication technologies applications – research programme

Another important governmental funding resource for the development of ICT applications in higher education is the basically research oriented IKTA

(Information and Communication Technologies Applications) a Programme of the Ministry of Education (Division of Research and Development). IKTA is made available in the framework of annual application procedures for consortia of higher education institutions and corporate or non-profit institutions. The annual budget is 4-6 MEUR and themes of the Calls include support for R&D, innovation, technology transfer, elaboration and trial of new, marketable tools, services and processes, development of ICT applications for public use, software and language technologies development, information services, information and knowledge management, and establishment of knowledge centres.

Evaluation of the implementation of IKTA underlines that there have been profound changes in the field of research and education in information technologies[4]. One decade ago, besides a number of research institutes of the Hungarian Academy of Sciences, national industrial research institutes existed and performed intensively. Research groups also worked at university departments. There were relatively few students in informatics and computer sciences, studying at only some universities and a limited number of individual PhD (doctoral) studies were offered. International cooperation was mostly based on personal acquaintance, and rather bilateral mobility agreements existed only with a limited number of countries.

189

The institutional background nowadays is quite different: a single research institute of the Academy of Sciences: SZTAKI (Budapest) remained on the scene, multinational companies and SMEs also entered the field. Research groups at (reorganised) university departments are increasingly active. There are naturally many students in informatics and computer sciences at several universities. New and old educational institutions accommodate numerous new departments of informatics. In the meantime, there is an aging staff working, very few juniors only and the age group of 30-50 is practically missing. The PhD (doctoral) studies are carried out in an organised way, with the involvement of a large number of students. Research cooperation is also running mostly in a coordinated way, with multilateral (mainly EU) partnerships, and mobility of researchers is also quite extensive.

The institutional background according to the evaluations does not correspond well to the increasing possibilities offered by the many calls for proposals in RTD and by extending international cooperation. However, there are outstanding results produced by a number of internationally well known Hungarian IT companies: Recognita – character recognition, Graphisoft –

ArchiCAD, MorphoLogic – language technologies, Cygron – data mining, Kürt Co – disk data rescue, SZTAKI – parallel and distributed computing, etc.

11.4 Pedagogy and ICTs – A Successful National Initiative on School Level: the 'Sulinet' Programme

An important barrier in developing ICT culture in schools was, even in the middle of the 1990, the lack of appropriate background in equipment and also the poor competence of teachers. Early in 1997 the large scale governmental development programme – 'SuliNet' ('SchoolNet') started. At that time, in over 8% of the institutions, there was no single computer and over 50% of the schools did not have any access to Internet. The Ministry of Education identified as a development priority the establishment of ICT infrastructure in the schools. In 1997-98, in the framework of the Secondary Schools' Internet Project with 7.5 MEUR central support, supplemented with 7.2 MEUR funds from local governments, every secondary school and nearly 250 grammar schools were provided with local computer network and Internet access. The development programme from May 1998 was continued under the name Írisz-SuliNet with partly changed aims and focus. The number of computer workstations exceeded 1800 and in March 2002, with the investment of another 13.6 MEUR, reached 2340. There was also another strand of technical development aimed at building up backbone lines, developing by March 2002 by ten times the capacity of the Hungarian network and broadening bandwidth to 34 MB from 17 MB[5].

Emphasis has shifted lately to the content provision, teacher training and bi-directional data traffic, and the establishment of the national information system in public education. Regarding content provision, the renewal of the Irisz-Sulinet home page www.irisz.sulinet.hu was the most important step. The number of daily visitors to the site nowadays is over 40,000. The impact of the SuliNet programme was further increased by the informatics, and specifically Internet courses, offered in the teacher in-service training system. In the National Core Curriculum (NCC) established for primary and secondary schools, informatics is identified as a separate subject, from the 5th year of studies. In the NCC, in the 5-6th year, 2-4% of the compulsory number of teaching hours (about 0.5 – 1 hours per week), in the 7-10th years of study, 4-7% (1-2 hours per week) have been recommended for teaching informatics.

A general difficulty in education has been the badly underfinanced status of institutions. Any momentous development is therefore heavily subject to governmental resources. Ensuring the conditions of maintenance and sustainibility of investments is problematic, as are personal conditions such as remuneration of teachers or technical personnel for the extra work requested.

11.5 Perspectives and Practice with the Use of Information Technologies at Universities

11.5.1 Survey of possible objectives and approaches

There are several possible objectives of the use of ICT in higher education: degree courses or continuing education for adult employees; use of ICT in on-campus, distance or correspondence education of full-time students; integration of outside education resources in the courses and offered by universities by relying on international and inter-university cooperations; broad introduction of ICT in mainstream education.

It is increasingly acknowledged that after long years characterised by a radical increase of student population and budget cuts, universities have face two new challenges: the implementation of the lifelong education concept and the extensive use of ICT in education. In this situation the majority of the institutions seem to be reluctant to build up new, comprehensive schemes for large-scale implementation of these new objectives.

In the field of continuing/lifelong education, universities have to enter the educational market, and compete with flexible and strong enterprises with a stable market position, tapping in several cases, the intellectual potential of higher education. It is essential therefore to apply market compatible structures, procedures and activities, which do not harmonise with the traditional university operation. Open and distance learning and ICTs may help to integrate the world of *academia*, with emphasis on content, scientific background, cultural and social context, development of critical thinking, morals, autonomy and freedom of teaching and learning, and the *business world*, with the keywords of competition, investment, division of labour, team work, profit, economy, marketing, sale, professionalism and management.

Concerning the question of how to integrate technology in the life of universities, according to Tamas Lajos (1998), three levels of application of ICT can be drawn up.

1   In the case of the *low profile approach*, individual lecturers *use the technology for enhancing effectiveness of the education.* Communication with students via e-mail, using the Internet for making course materials and literature available are typical applications. This approach fits well into the traditional academic activities; it does not require considerable additional investment. Universities can promote this activity by stimulation and training of academic staff.

2   The *medium profile approach* means the *systematic integration of the technology* in the life (not only in the education) of the institution. This already needs considerable investment and operational costs. A top-down approach, strong technical support, standardisation, extensive staff development, planned and structured teamwork are the key elements. These features partly conflict with traditional academic activities, which are based on individual initiatives and achievements, integration of research and education, and academic freedom. This conflict can be reduced by reasonable division of labour, where specialised new units of the university take over the bulk of the development work in close cooperation with the academic staff.

3   The third *high profile approach* is *the implementation of the virtual university concept.* The same high profile approach is needed, at least in a part of the institutions, if the university strongly enters the continuing education market. In this case, because of the fundamental change in operation needed at a university, which can be characterised by comparing the work of artisans and industrial production, not only large-scale investments are needed, but also deep transformation of university structures, life and partly missions. Establishment of relevant new units and cooperation with outside institutions may reduce the degree of necessary changes.

In international comparison, the high profile approach can mostly be found at US universities and at a few European flagship institutions (open universities). Otherwise, with big differences between countries and types of institutions, low or blended medium-low profile solutions can be observed.

Regarding the *present situation in Hungary* the application of ICT in on-campus education of full-time students in order to improve teaching and learning through the low profile approach is characteristic. The achievements of the earlier quoted national development programmes are also appearing but so far rather as isolated examples. There is a problem with financing and sustaining the developments: the critical mass of students is missing, which could ensure the sustainability of activities and justify the investments for the larger scale course developments.

11.6 Current Trends in Higher Education in Hungary

The significant phenomena in Hungary in the higher education sphere during recent years have been the reduction in number of institutions and the continuous increase in student numbers. In 1990, there were 77, in 1995 there were already 90 higher education institutions in the country. From 1996, the integration of the state-owned institutions started, resulting by January 2002 in only 67 higher education institutions, out of which 31 were functioning with state support, 26 were maintained by churches and 10 by non-profit institutions.

The increase in student numbers can be traced back to the fact that between 1990 and 1994 the number of youngsters entering in their 18th year increased by about one third. However, since then, this ratio has decreased again: the number of full-time students shows a constant increase of 6% per year. The increase in overall student numbers can be explained with the development of student population in non-full-time education, particularly in distance learning.

As a new experience, there are more people in the relatively older age groups applying for full-time studies. At the end of the 1990s, three times more 26-29 year old persons asked for admission to universities and colleges than at the beginning of the decade. This shows also the acknowledgement of importance of education and the increased value of the degree earned.

193

An important element of the expansion of higher education in Hungary is the proliferation of the courses offered in distance education form. In 2002, more than 20% of the higher education institutions, altogether 15 universities and colleges, offered 37 majors for undergraduate distance education. In the academic year 2001/2002, there were 349,000 students in higher education; out of these, 193,000 were full-time. The number of full-time students increased by 5% whilst that of evening, correspondence and distance education courses increased by 9% in comparison with the previous year. About ten percent of the students attended distance education courses. With one exception, the distance studies offered were at college and not university level. The only institution offering fully Internet based e-learning is the Pázmány Péter Catholic University which in the field of law studies provides such education leading to degree.

Development and widespread use of ODL and ICT-based education in higher education has resulted in the necessity of accreditation of these forms of education, which are different from the traditional delivery modes. In accordance with the Act on Higher Education, the Hungarian Accreditation Committee is reviewing in 2002 the off-campus and distance education courses.

In informatics, the number of undergraduate students has considerably increased: in 2002, 25 universities and colleges (37% of all institutions) offered 100 majors separately or in combination with other majors[6].

11.7 e-Learning in the Higher Education and Corporate Sector: Comparison of Priorities

According to a recent survey (Eduweb Co, 2001), in *the corporate sector* in Hungary, the order of preferences regarding the advantages of ODL and eLearning were the following:
1   Faster learning;
2   Efficient skills development;
3   Efficient knowledge transfer;
4   Lower costs;
5   Preference of learners to learn at home;
6   Flexible timing for the learner.

Clear preference was found for the blended type of learning (classroom based, computer based, web based). Expectations show that in 4-5 years time the importance of electronic learning may overtake the learning using traditional (printed) media.

The main tasks emerging and problems mentioned in *the higher education sphere* regarding the introduction of e-Learning and distance education were the following:
-   difficulties with the integration of such educational programmes in the daily operation of the institutions;
-   lack of student knowledge about how to learn;
-   access to and competence in using IT equipment for education;
-   limited digital literacy of teachers;
-   professional course development, including training of university teachers in developing eLearning/ODL courses;
-   need for 'positive discrimination' of courses in governmental support.

Corporate sector and educational sector respondents agreed about the utmost importance of supporting the development of e-Learning and ODL courses, the support for the development of methodologies and the evaluation and accreditation of the courses and programmes. Concerning the question asking which activities or fields would need most urgent financial support, respondents ranked in first place the development of content: e-Learning/distance education courses, and only with smaller preference infrastructure developments or even for increasing the personal costs for experts working in the institutions.

Forecasts state that within the next 4-5 years the computer-based training in about 75% of the institutions will play a significant role. This indicates a high predicted development rate, particularly expected in the field of web-based education methods. Such potential development could ensure a serious background for the development of electronic learning, particularly if the governmental funding and strategy were available.

11.8 The Institutional Response – A Case Study from the Budapest University of Technology and Economics

195

11.8.1 A survey on separtment web sites for education

In 2001, a survey was conducted in the Budapest University of Technology and Economics, carried out by its Distance Education Centre, to analyse the e-learning potential of the University. The survey was based on investigation of the public web sites of all University departments, using an extensive list of qualitative and quantitative criteria. (BUTE Distance Education Centre, 2001).

It was found that practically all departments have their own home pages, although they are very diverse in structure, design, quality and content. 97% of the departments that own home pages have general contact e-mail addresses; furthermore almost half of them (48%) provide access to contact data of the teaching staff.

The web pages are also functioning as a resource for information about the courses offered. Most departments (72%) post their course lists on their sites, 64% supplement these lists with full or partial course descriptions, but only 9% give detailed lecture syllabus. The electronic administration of mid-term grades and test results is used by 9% of the departments. A widespread use of the Internet as an information source for higher education is the publication of lecture notes. 42% of the departments take this opportunity either by uploading various files, or transforming the materials to a format that can be viewed by the web browsers directly. 41% of the sites also offer additional link collections to help the students to find additional sources of course-related information.

To help understanding of the course material and to ease the studying process a relatively large number of departments (18%) provide access to helpful softwares and animations on their sites. To make the examination expectations clear, 33% of the departments publish earlier examination themes, tests and their keys.

Concerning the sites' interactivity, the analysis provided reasonably good results: 16% of the home pages have either guest books and forums for off-line conversations or chat-rooms for on-line discussions. It should be added that the University runs a comprehensive, interactive student administration and information system (NEPTUN) which allows students to arrange most formalities related to the administration and management of their studies from any Internet connected PC at the campus or even from their homes or dormitory rooms. All university units and departments are surveying their financial affairs through the institutional electronic network.

11.8.2 The central ICT and education linked units

Besides the Faculty of Electrical Engineering and Informatics, the University has established a *Center of Information Technology*, in support of the integration, methodological support and educational technology modernisation of informatics education within the University. The Centre is expected to mobilise external support and cooperation with the corporate sector for these purposes and promote cooperation with the main players of the IT industry and economy. As a result of this effort, a number of support and cooperation agreements have been concluded with leading corporations in the IT field like Compaq, which, besides valuable hardware, offers research scholarships for professors and students, or IBM which established an eBusiness Academy at the University[7].

In 1997, the University established its *Distance Education Centre*, working with 10 full-time staff, including experts in informatics and ODL technology, course developers, training coordinators and project coordinators. The Open and Distance Education Laboratory, with over 20 PhD students, works closely with the Centre, creating solid academic and professional background to the activities. The contribution of the Centre to Hungarian governmental projects in 1997-2000 in developing pedagogical standards and quality management systems of national networks served as a basis for the implementation of large scale initiatives at national level. The Centre has been involved in several EU projects and hosts one of the Study Centres of the EU PHARE Multi-Country Distance Education Programmes. There are a number of assignments, successfully accomplished by the Centre, for Hungarian national and multinational companies and public institutions in the fields of educational system development, course development and consultancy[8].

Besides standard audiovisual facilities, the Centre is equipped with video-conferencing equipment, multi-media PCs, linked to the optical fibre ring providing wide bandwidth to the universities of Budapest and the outside world. Part of the activities is the multimedia evaluation and development laboratory, running in cooperation with the Department of Ergonomy. The 'Support for Visually Impaired' Unit of the University is also accommodated in the Distance Education centre and receives IT and methodology support for their activity. The Distance Education Centre, like the Center of Information Technology, is performing without financial support from the University, on the basis of assignments, service contracts and projects. The approach needed for efficient functioning includes systematic identification and exploration of business opportunities, the combination of different funding possibilities and development of public-private partnerships. This means in the meantime that the professional support delivered for the university teaching itself is smaller than it could and should be.

On a valuable building site, near the campus of the University, the 'Infopark' *Information Technology Park* has been established, already accommodating research and development divisions of a number of leading international IT corporations like IBM, Hewlett-Packard, Pan, and the Hungarian Telecom. The Infopark Company is co-owned by the University and an international real estate agency and the Faculties offer their research capacities and services for the companies in Infopark[9].

11.9 A Look at Central and Eastern Europe

Heads of Government of the EU Candidate Countries agreed in June 2001 on common action to address the challenges of the Digital Economy by launching the *eEurope+* Action Plan issued on the occasion of the Göteborg Summit[10]. A common Action Plan was presented to the Heads of State and Government of the EU, to underline the Candidate Countries' full subscription to the ambitious goal agreed by the EU leaders at the Lisbon Summit about eEurope in 2000.

The eEurope+ Action Plan aims to help in accelerating reform and modernisation of the economies in the Candidate Countries, encourage capacity and institution building, improve overall competitiveness, and allow the countries to apply their strengths to the advantage of the citizens.

The Action Plan is based on four objectives: (i) to accelerate the setting up of the basic building blocks of the Information Society, (ii) a cheaper, faster, secure Internet, (iii) investing in people and skills, and (iv) stimulating the use of the Internet, and also a whole range of actions in areas like e-commerce, education, e-health, e-government, transport, and environment. Candidate Countries expect to call for assistance from the private sector, the international financial institutions, the not-for-profit sector, and the social partners in the implementation of the plan, and anticipate substantial related private sector investment.

According to recent information, in July 2002, the Hungarian Minister of Informatics initiated formally at the European Commission – together with the Ministers of Slovenia and Poland – in support of the upcoming EU accession, the offer of joining the eEurope programme for the candidate countries. It was expected that a positive reply will be given by the European Commission[11].

11.10 An EU Regional Initiative – Experience with the Phare Multi-Country Distance Education Project in Central and Eastern Europe

Although theoretical and practical knowledge and experience of second generation distance education methods have been available for decades, European education systems (with few exceptions) did not react in a relevant way, and open and distance learning could not move from the periphery of the human resource development sphere. In the 1990s, the situation has considerably changed: the development of ODL moved up on the priority list of most regional or national human resource development strategies and significant programmes of the European Union started to support them substantially. The main reasons for the favourable change was the rapid development and expansion of the use of information and communication technology. In education, the efficient use of ICT requires the adoption of the underlying principles and methods of ODL. The shift of open and distance learning towards the centre of human resource development is not mainly the consequence of development of distance education as a methodology itself, but rather due to the change of regional and national education policies, reacting to the rapid technological changes and the concept of the information society (Lajos *et al.*, 2000).

11.10.1 *An attempt for regional response*

In 1994, with the support of the EU Phare programme, the European Commission launched a Regional pilot project in distance education in the Phare actions for Regional cooperation.

The key objectives of the project included ICT infrastructure investment supporting the establishment of national contact points and study centres in the beneficiary countries, in order to develop the conditions which allow the participating countries to interact and cooperate on an equal basis, and also to act as a catalyst for national policy formulation in the field of distance education and to define areas of common interest in which regional cooperation can support national policies, in terms of enhanced quality of the outputs, speed of development and/or economies of scale.

In 1995-97, National Centres for Distance Education (NCDE) and National Contact Points (NCP) were established, trans-national and national awareness raising and train-the-trainers programmes implemented and pilot courses developed. On the basis of the achievements of the Pilot Project the Commission launched in 1995 for all Phare partner countries, a follow-up programme resulted in 1996-99 in the continuation and consolidation of the NCDE-NCP network and the establishment and equipping of 40 ODL study centres and human resource development for the project. Local and regional course development programmes have been implemented, further strategy studies were elaborated on infrastructure and market development issues in ODL in the region and about legislation, recognition and quality assurance methods.

The presence of the European Union in the establishment of this structure has offered a special dimension for the above developments. As usual, the financial resources ensured by the EU Commission from the Phare budget were not huge; however, the final output of the whole operation, strengthened by the coordinating presence of the European dimension and by the additionally mobilised national resources has ensured an impressive order of magnitude of achievements (Szucs & Jenkins, 1999).

11.10.2 Evaluation of experience

The experience of this large-scale programme has been analysed and investigated by a number of experts and their groups, from different angles. A detailed evaluation of the programme has been carried out by EU experts and a summary of that work published (Mackeogh & Baumeister, 2000). They found that '... the specific objectives of the programme had been met ... within educational systems which were themselves in transition during the four years of the programme. The design of the programme, based on a networking approach involving a wide range of institutions and individuals both from the education sector and the political sector had worked well.

> (...) Barriers to embedding odl in the national structures are largely financial, and environmental. The Phare programme demonstrated that it was possible, for a relatively small investment, to build a potentially sustainable infrastructure, involving national contact points, a network of some forty study centres, development of course modules, and a range of activities designed to lay the foundations for a viable distance education system in each country.' (Mackeogh & Baumeister, 2000).

The recommendations of MacKeogh and Baumeister included facilitation the sustainable introduction of the new approaches. The networks established should be maintained and expanded on a regional, national and trans-national level. Legislation and infrastructure should be established to enable further development to take place. The focus of effort should be on sustaining and enhancing the human resource base which has been developed by the programme. However, since the programme funding stopped at the end of 1999, the future has been unclear. There are no plans to continue the programme. This is an illustration of the difficulties in implementing change, using top-down, short-term programmes.

11.11 Expectations and Realities

Before the political and economic transitions at the end of the 1980s, the previous monolithic political and economic structure in most countries of Eastern Europe only needed and appreciated the autonomous and intelligent workforce and personalities in a limited way. Unfortunately, at least until the turn of the millennium, the new economic system – either the multinational companies expanding in the region, or the proliferating but 'just-surviving' SMEs – did not seem to require this type of input. The poor financing ability and willingness of the public sector and governments, and the reserved position and different preferences of the corporate sector has not enabled the momentous development of open and flexible educational systems so far.

Public sector and governments are facing the challenge of reaching a certain critical mass, either as initiator of determining processes or providing an appropriate financial input or institutional inertia for the processes. The development of an open, efficient, flexible training system could certainly carry a strong modernisation potential which, combined with the catchword of the information society, would possibly be able to play an important role in the development of the region.

201

The turning point is still ahead and it remains a crucial question, to what extent (and in which fields) the economies and societies in the countries of the region will request the inputs and services of modern educational methods, and how e-Learning can respond to the needs emerging, attract funds, and resources either from governments (labour market development) or from the corporate sector. In this respect, quite different situations, country by country can emerge and the economic position of the countries, the extent to which they can mobilise companies, requesting higher knowledge and innovation content will be a determining factor. A further essential question (particularly in the light of the eEurope initiative) will be, how far comprehensive national infrastructure development programmes will support the establishment of open learning systems.

References

BUTE Distance Education Centre. (2001). *Survey on the 'E-learning potential' of the Budapest University of Technology and Economics based on the investigation of the public web sites*. Budapest: BUTE.

Consulting & Research for Industrial Economics Ltd. (2001). *Infocommunication sup-port of education, training, lifelong learning in Hungary Virtual-Space Observatory.* Budapest: Office for Government Commissioner in Charge of Information Technology. (in Hungarian). http://www.ikb.hu/c3m511.html

Eduweb Co. (2001). *Position, perspectives and trends of development of eLearning in Hungary.* Budapest: Eduweb Co. (in Hungarian). www.eduweb.hu

Lajos, Tamas. (1998). From Traditional to Virtual: New Information Technologies in Education. In: *Proceedings of the* UNESCO *World Conference on Higher Education,* Paris, 5-9 October 1998. http://www.unesco.org/education/wche/pdf/th_new-tech.pdf

Lajos, T., Grementieri, V., & Szucs, A. (2000). odl networking in Europe and the expe-rience of the East-West cooperation. In: Armando Rocha Trindade (ed.), *New Learning.* Lisbon: Universidade Aberta.

Mackeogh, K. & Baumeister, H., P. (2000). Policies and Practice in Open and Distance Learning: The Phare Multi-Country Programme for Distance Education 1995-1999. Book of Essays of The First Research Workshop of Eden: Research And Innovation In Open And Distance Learning, Prague, 16-17 March 2000. EDEN – European Distance Education Network.

Office for Government Commissioner in charge of Information Technology. (2001). *National Information Society Strategy.* Budapest: Office for Government Commissioner in charge of Information Technology. (in Hungarian). http://www.ikb.hu/download/nits10.pdf

Prime Minister's Office. (2000). *Theses about the Information Society.* Budapest: Prime Minister's Office. (in Hungarian). http://www.kancellaria.gov.hu/tevekenyseg/kiad-vanyok/tezis.htm

SZUCS, András and Jenkins, Janet. (1999). Distance Education in Central and Eastern Europe. In: Keith Harry (ed), *Higher Education through Open and Distance Learning.* The Commonwealth of Learning Routledge, London..

UNESCO Information Society and Trend Research Institute. (2000). *Hungary's Information Society Development – preliminary study.* Budapest: UNESCO Information Society and Trend Research Institute. (In Hungarian). http://www.ittk.hu/infinit/2000/0420/

Notes

1   Governmental Portal, http://www.ekormanyzat.hu/
2   The Széchenyi Plan, Ministry of Economics,

http://www.gm.hu/gm/alolda.htm?d=6&k=http://www.gm.hu/kulfold/english/ang ol/o.htm

3   Apertus Public Foundation, http://www.apertus.hu

4   Snapshot of Hungarian R&D activities in Information Technology – Dilemmas of Subsidies, Peter Hanak, Ministry of Education, Division of Research and Development, Department of Technologies of the Information Society, Peter.Hanak@om.gov.hu

5   English homepage of the Sulinet programme, http://www.sulinet.hu/cgi-bin/db2www/lm/kat/lst?kat=be

6   Home page of the National Office for University Admission, http://www.felvi.hu

7   Budapest University of Technology and Economics, Center of Information Technology, http://www.ik.bme.hu/indexe.html

8   Budapest University of Technology and Economics, Distance Education Centre, www.bme-tk.bme.hu

9   Infopark, www.infopark.hu

10 Commission welcomes eEurope+ initiative of EU Candidate Countries, http://europa.eu.int/comm/gothenburg_council/eeurope_en.htm

11 http://index.hu/tech/ihirek/?main:2002.07.12&97372

12 The French Approach to e-Learning: Public Initiatives for Virtual Campuses and a Francophone Medical School

Anne Auban, University Pierre & Marie Curie, France
Bernard Loing, International Council for Open and Distance Education at UNESCO

12.1 France: a Latecomer in e-Learning

Like people, countries have their own idiosyncratic way of evolving, of following their timeline. Some grow smoothly and evolve peacefully. Others have a more bumpy course, with sleepy periods, glorious peaks, catastrophic revolutions. France is rather of the latter type, not only in its political history, but also in the history of its education.

Without going too far back into the past, we can see that after the upheavals of the student movements of 1968, French higher education had settled down into a quieter period, in which its 91 universities (including several new ones), plus a few hundred various higher education institutions (including our celebrated 'Grandes Ecoles') were developing on their own grounds, not bothering too much about what was going on in the outside world. And when the new bandwagon of e-learning started rumbling on, along the Internet highway, nobody, apart from a few eccentric academics, seemed to care very much within French universities: the 'teaching business' was going on as usual in class rooms and amphitheatres.

Compared with what was happening in other countries at similar levels of development, both European and Anglo-Saxon, France was obviously lagging behind in the educational applications of ICT at higher education level. (This was probably less true at secondary level, but this is another story, which we are not going to tell today.)

The main reasons are easy to list:

- The total number of universities, institutions and other higher education facilities is large enough in France to cope with the total number of students, i.e. roughly 500 for 2.16 million students, which gives an average number of 4300 students for each establishment[1];

- France has no 'Open University', CNED being a national distance learning facility for all levels of study, from primary to higher education and lifelong learning; university degrees are delivered by partner universities. As such it covers 80% of the nation's needs in distance education and open learning, with the advantage, because of its different status, of avoiding frontal competition with regular universities;

- As everywhere, but maybe more resolutely, there has been some resistance of academic staff to the use of ICT for teaching, if less so for learning;

- Supposed to be 'autonomous, a rather questionable qualification since they are all financed on public funds, each university tends to live in relative isolation, reluctant either to join, or to compete, with others, whereas partnership and competition are, as it were, prerequisites in the transnational open field of e-learning.

- Above all, there is the difficulty of finding adequate resources to implement the educational use of ICT: Human resources are rare in the field, from on-line tutors to webmasters and experts in educational software, and academic careers have yet to be adapted to include such professions. Financial resources (especially coming from public funding) are most of the time inadequate, if one considers for instance, that one hour's technology-based course costs from 4 000 Euro to 12 000 Euro, depending on the discipline and on the level of study.

Thus, for all those reasons, and probably a few others, France was lagging behind although its technology, especially in telecommunication, was at least as good as any other. Indeed here and there, initiatives had been taken by individual or by groups of universities, to implement some sort of multi-media, on-line courses that could be accessed at a distance. In the year 2000, one can estimate that various combinations of learning platforms had been established in some 30 universities, with specialised rooms, on-line services to students, e-mail addresses, tele-conferencing, sometimes even on-line tutoring. The disciplines concerned were mostly science, medical studies, management, engineering; and less often language learning, law, geography and history.

Among the more ambitious systems existing then were, to mention a few of them:

- RUCA (Réseau Universitaire d'Autoformation) created as early as 1987, a network of resources for the self-training of students, including 15 university partners;

- AudioSup Net, an Internet portal created five years ago, offering thousands of digitised courses delivered by radio in some 25 universities over the last ten years;
- GRECO (Grenoble Campus Ouvert), a recent initiative taken by the four universities of Grenoble.
- CNED the National Centre for Distance Education, which was already developing its own 'digital campus';
- And in the field of professional training, a portal called On-line FormaPro offering some 200 modularised courses some of which including tutoring.

12.2 The Digital Campuses: A Major Governmental Initiative

Yet all those efforts were relatively scattered, lacking coordination and showing no common strategy. Confronted with the growing competition threatening from abroad, made particularly obvious by the first 'World Education Market' held in Vancouver in April 2000, the French government decided to react.

Coming after a general agreement had been signed between the Conference of University Presidents, CNED and FIED (the Federation of universities already engaged in distance education operations), *an official call for proposals was launched in June 2000 to create a national network of 'Digital Campuses'*. It was addressed to all potential participants in the field (universities, higher education institutions, CNED, etc.) and also to companies and corporate businesses, in a spirit of open partnership. This first call for tenders was followed by two similar calls, one in 2001 and another one this year, which has just been closed.

Although it is as yet too early for an evaluation, a first tentative overview of the situation, rather positive, of the Digital Campus Project can be sketched as it now stands in mid-2002.

Conceived and organised as a major strategic move in favour of a systematic integration of ICT into the French education system, the Project raised from the start considerable interest among the higher education community. It was an incentive for all; and for those universities, especially the multi-campus ones, that had already started laying out their own strategy for an integration of ICT into their teaching practice, the call was coming as a boost for implementation. Thus it mobilised some 400 partners among which there were 194 higher education institutions in France, 35 abroad (of which 13 European and 20 in francophone countries), and 154 various companies, businesses and as-sociations both in France and abroad. CNED is present in 19 of the projects as an operator.

12.3 Goals, Principles and Issues

The term 'digital campus' may be a little misleading to describe the project. Indeed, the Digital Campuses are anything but new universities. They should rather be described as consortia of facilities in a digital mode to provide easy and flexible access to high quality educational resources provided by the partner institutions, within a modular validation system of the ECTS type. But even if the main objective was not to create new higher education institutions, the strategic goals of the project were very strongly and clearly defined, as the call for proposals was launched. They were:
- to keep up with the emerging Information Society at global level and prepare for international competition
- to cater for new users especially for those engaging in lifelong learning, and offer new learning services to students, both on campus and at a distance;
- to create powerful tools and find new ways of teaching and learning
- to be proactive in the setting up of a European system of accreditation (ECTS).

The basic principles on which those so-called 'digital campuses' are founded are:
- an institutional and mutualised approach for the use of ICT in higher education;
- a logic of tender and open partnership to build up economic incentives for the universities;
- a priority given to digital technology to provide innovative services for students;
- a double policy line: 1. ODL systems for distant students, for lifelong learning, and for international contacts and influence. 2. New learning environments on campuses for all students.

From the start, concrete and important issues were raised which are common to all initiatives of that kind:
- What would be the status of the teaching staff engaged in ICT activities? And how could those activities be integrated into regular academic duties and career?

– Legal questions also had to be addressed: what sort of 'pedagogic fair use' could be set up as a limit to copyright? Another relevant question in that field concerned the legal and formal technicalities of the consortia framework which sometimes seem somewhat loose;
– What quality specifications ought to be defined and applied to the hardware and software provided (networks and student equipment)?
– What overall quality assessment should be applied to the digital campuses, and by whom?
– How could the whole system become sustainable?

12.4 Digital Campuses: Resources, Programmes, Curricula

Supported by the Ministry of Education, the project is funded on specific public investment:
– for the year 2000: 18 million francs = 2.74 MEuro;
– for the year 2001: 61,5 million francs = 9.38 M Euro;
– for the year 2002: 66.5 million francs = 10.14 M Euro.

One can immediately see that those sums are relatively small for such an ambitious project. But they are rather given as an incentive for the first year since the project must be self-sustaining from the second year onward. Universities can also draw on two other sources of funding:
– The funds specifically provided by the Ministry of Education within the 4-year contract passed with each university for the development of on-line learning (in 2001, the national amounts were 30 million ff to develop courses, and 15 million ff for local infrastructures);
– The funds provided by the local authorities, which in some regions can be quite generous.

In a first report published on the project in April this year (Rapport Averous/Touzot) the overall necessary investment at national level was estimated at 150 Mff a year, plus 30 Mff specifically devoted to the yearly creation of on-line courses. Such a budget means that a substantial amount of money will have to come from private sources, usually coming from continuous education and professional training; which means that a significant part of the Digital Campus Project will have to be devoted to such activities if it wants to be self-sustaining.
So far, 72 projects have been supported (45 of which are already open and active), for different disciplines and curricula, with a majority of scientific subjects:

– Science and technology: 23 (15 + 8);
– Health and medical studies: 15 (9 + 6);
– ICT and educational engineering: 13 (8 + 5);
– Economy and management: 8 (6 + 2);
– Arts and social sciences: 7 (4 + 3);
– Law: 5 (4 + 1);
– Documentation: 1.

As regards the pedagogic engineering of the whole Project, it consists of various pedagogic resources made available, mostly on-line, and of actions of support by asynchronous tutoring and collaborative work.

An overall survey of the pedagogic resources available shows that:
– 60% are multi-media;
– 38% are hypertext;
– 2% are textual.
New tools are required, which hopefully should be available in the near future, such as digital text books and on-line university libraries.

As for the location of the resources offered, their situation shows a priority given to distance education, and to the extension of an international influence, at least in the francophone world:
– 71% are available both on campus and at distance;
– 24% are available only at distance;
– 5% are available only on campus.

– 59% are located both in France and abroad;
– 32% are located only in France;
– 9% are located only abroad.

The types of curricula proposed in 2002-2003 clearly show that the Digital Campus Project aims at covering the whole academic range of the higher education system. They also illustrate the attention given to professional training and to the implementation of a lifelong learning system in universities.
Out of a total number of 134 courses proposed:
– 33 concern only academic initial education, among which 23 have a full curriculum leading to the higher degree of Doctorat d'Etat, 9 have a shorter curriculum leading to the lesser degree of Doctorat d'Université, and 1 has no final degree;

- 34 concern both initial education and professional training, among which
  19 lead to a Doctorat d'Etat, 8 lead to a Doctorat d'Université, and 7 have no
  final degree;
- 67 concern only continuing education and professional training, the
  majority of which (48) lead to no final degree.

As for the number of students registered in those courses, the following
figures can be given:
- 2, 119 for the year 2000-2001;
- 6, 185 for the year 2001-2002;
- 23, 300 (estimated) for the year 2002-2003.
It is generally considered that this type of tuition should in the end concern
about 2% of the French student population, that is, about 40, 000 students.

The 2002 call for proposals provides for a new development in the Project
(under the name of 'Phase 2'), 1.52 Mff being reserved for the creation of new
digital learning environments located on campus and offering various services
to students and staff.

12.5 Case Study: The French Virtual Medical University: UMVF

211

The creation of UMVF is a French initiative to learn and teach Medicine and
Surgery, coordinated by the University Paris 6. This project was launched in
1999 with the support of the French Ministry of research and Development. It
has the agreement of the Conference of French Medical-School Deans, in
conjunction with the reform of medical studies in France. The UMVF is a
consortium of 24 universities and industrial companies, the CNES is one of
them, under the supervision of the French Ministries of Research, Education
and Health. The UMVF is not only a portal (Figure 12.1). It has to provide to its
partners services for their professors, students, researchers and patrons. The
aim is also to share common resources, specific tools and developments. It will
integrate, in the end, most of the other 'Digital Campuses', supported by the
Ministry of Education, for specialities like Microbiology, Emergencies,
Radiology, Gynaecology-obstetrics, Microsurgery, Genetics, ORL, Odontology,
and other medical specialisms (Figure 12.2).

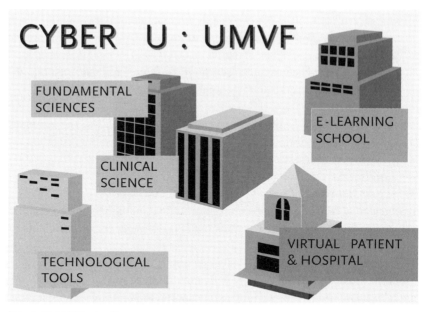

Figure 12.1 UMVF structure

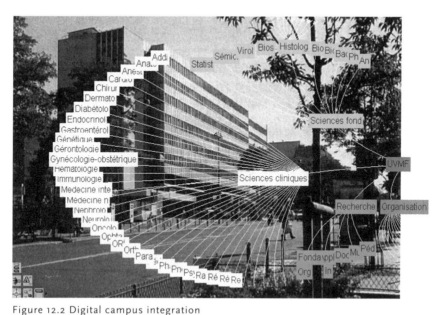

Figure 12.2 Digital campus integration

12.5.1 ENELM: how to integrate multimedia systems in pedagogic methods

Creating resources for teaching medicine on the Internet will require specific efforts for the teachers and the faculty. Furthermore, new education programmes, based on problem-based learning, are required by the new French medical curriculum. This was the reason why it was obvious that it is fundamental to create a national school to teach medicine teachers how to integrate multimedia systems in these pedagogic methods.

The E -Learning Medical National School, so-called ENELM -'Ecole. Nationale de E-Learning pour la Médecine', is one of the digital campus supported by the Ministry of Education. Specific multimedia technologies and tools are in development to enhance the disciplinary contents in fundamental and clinical sciences. The project is aimed at creating a database of a French, national and francophone, academic knowledge in health and medical education fields. The speciality collegiums and the universities mainly provided the data: tutoring, e-Learning, multimedia courses, text files, and PowerPoint presentations.

We can list specific goals using information and communication technology such as:
− Offering and testing on-line audio and video multicast and streaming for live or recorded training resources;
− Conceiving specific tools to index the resources using MeSH thesaurus and UMLS meta-thesaurus;
− Conceptual modelling to seek contents in a contextual way.

One of the main issues is to know how multi-media systems can be integrated in these new pedagogic methods used now in medical studies; for instance, how to develop various teaching approaches using the contents of clinical and factual databases, how to take into account the scientific results of Evidence Based Medicine in diagnostic and/or therapeutic decision-making.

12.5.2 Medical e-learning is a hybrid concept mixing reality and virtuality

Medical studies as well as formal education for vocational training need to improve professional competencies. The disciplinary field is broadband, from theoretical topics to experimental or clinical ones. To support new pedagogical approaches, like problem-based learning in a context of Open and Distance Learning (ODL), it is very important to create hybrid models with interfaces between teachers (real or virtual), students (real or virtual) and patients (real or virtual) at the Medical Schools and hospital (real or virtual). See Figure 12.3.

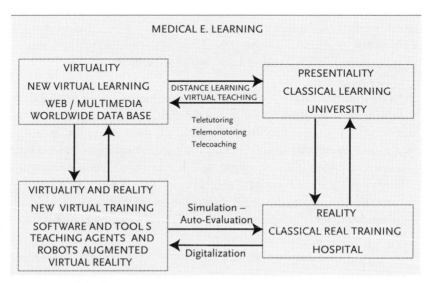

Figure 12.3 A hybrid concept

Mixing reality and virtuality needs to work and study on real but also virtual patients and hospitals. The pictures in Figure 12.4 (screen computers hard copies) show significantly different mixed situations described in Figure 12.3. In the same pedagogic context we can have virtual and real participants. They also show how the use of digital technologies, like video streaming, video-conferences, simulation and interactive multimedia courses can be integrated in the formations.

It can be concluded that creating resources for teaching on the Internet will require specific efforts for the teachers and the faculty and needs a strong political will for priority funding for the scheme. But the future is today and as Alan Kay said: 'One of the best ways of predicting the future is to invent it. This is the century, which, if you have a good vision, you can actually build it.'

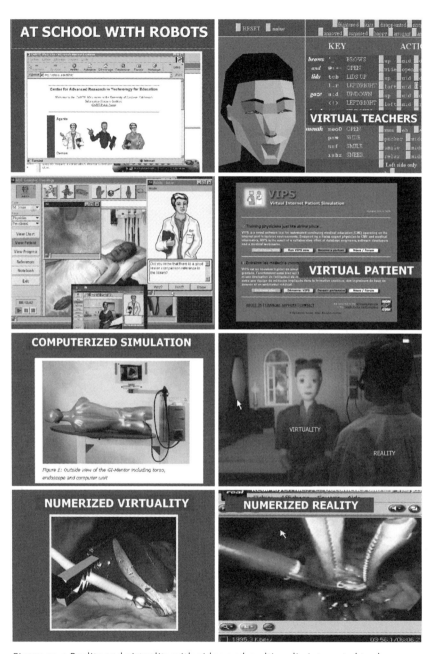

Figure 12.4 Reality and virtuality with video and multimedia integrated tools

12.6 Conclusion: a Collective Effort; New Regional Trends; the Sustainability Problem

As a conclusion, one can say that this major governmental initiative has certainly stirred considerable interest among the higher education community in France, now fully conscious that its future is at stake. It has also mobilised thousands of students and staff; in a powerful contribution to awaken the university world to the potential offered by the new technologies. Thanks to the project, a new spirit of open partnership is now prevailing, within the higher education community as well as between universities and the corporate world.

Beyond the systematic introduction of new ICT tools for teaching and learning, one very positive, if not totally expected, side effect of the project is the influence it bears on a new regional structuring of the French higher education system. To answer the call for proposals, a majority of universities have found it convenient to organise their partnerships in regional clusters. What had been done previously by the four universities of the Grenoble area with the GRECO, has now spread to several regions around major cities like Bordeaux, Toulouse, Montpellier, Rennes, etc. inducing a new kind of collaboration among universities that are quite near to each other, but most of the time working in relative isolation.

The major uncertain issue remains that of the *sustainability* of the whole system, which mostly depends on its financing. As mentioned above, it is clear that the initial public funding is provided to give the initial impetus, but remains largely inadequate to lead the project to its maturity. An economic balance will have to be found between the logic of a commercial offer usually linked to professional training, and the logic of a public service which is rather that of the French traditional academic system.

The recommendations of the Averous-Touzot Report (April 2002), our main source for this presentation together with documentation coming from the French Ministry and various partners in the project, are that the Government should implement a specific mechanism to finance adequately (i.e. corresponding to the above-mentioned needs of 180 Mff a year) those digital campuses able to show their potential for development, at least for an initial period of 4 to 5 years. In any case, even if a substantial part of the financing could come from private sources via professional training, it seems obvious that the share of public financing, both national and local, should remain fairly

high, considering the high priorities given to such issues as:
- the development of lifelong learning and access given to new users;
- the European action plan in favour of e-learning;
- the cultural and economic support aimed at francophone countries;
- the position to maintain in the global competition.

As for the campuses, in order to keep quality at a lower cost, they should:
- industrialise and automate the production of multi-media digital documents as much as possible;
- outsource the technical implementation of the software;
- concentrate the expertise of their available staff on content and pedagogy;
- open up large partnerships for the development and diffusion of their products.

Notes

1   The French higher education official figures for 2001-2002 are: 91 universities, 5 Ecoles Normales Supérieures', 19 private higher education institutions, 153 enginee-ring schools and colleges, 234 business and management schools and colleges. The total number of students for the same year was 2 159 556 (among which 1, 404, 014 registered in the universities).

13 The Approach of Italian Universities to ICT for Teaching and Learning: Contextual Elements, Achievements and Future Prospects

Claudio Dondi, Scienter, Italy

13.1 The Context: Infrastructure and Policy

Amongst the four largest economies of Europe, Italy is certainly the least advanced in the penetration and use of ICT, although rapid progress has been observed in the last three years. Recent researches[1] conducted on the ICT market show a stable trend on IT expenditure accompanied by decreased company investment and increasing variety of IT-related competencies demanded. The skill shortage amounted to 85,000 personnel in 2001 and is estimated to be around 32,500 missing professional profiles, leading to a 'missed market' of more than 3 billion Euro in 2002. In this context, the number of bottom-up initiatives promoted by higher education institutions is increasing, despite the fact that the tradition of distance education at university level is not so long as in most other European countries. In fact, only in the last decade a major national initiative, the Consorzio Nettuno (www.uninettuno.it), has developed as a reference model for academic cooperation in the provision of diploma courses, recently also degrees, via satellite tv and the Internet.

Within the Italian higher education system, which includes 64 Universities and 298 sites all over the territory and serves 1,658,000 students (in 2001), other national projects and initiatives are in place, whilst the situation is one in which a multitude of experiments of use of ICT in teaching is progressively generating institutional offers by single universities, with a general feeling that a national policy would benefit the credibility of e-Learning supply by public and private universities.

13.2 Higher Education and Research

The increase of bottom-up initiatives promoted by Italian Universities (often in cooperation with private participants) is probably the main result of the process of autonomisation featuring the education sector in recent years.

The role of the Government in the innovation of the higher education system should, however, not be underestimated: in the last 2 years, the government has undertaken various initiatives in order to:
– reform the Higher Education system;
– integrate ICT in higher education didactic (implying new competencies for teachers, and the introduction of IT courses in all University courses);
– endow universities and research centres with the necessary IT infrastructure; and
– intensify the relationship between universities and industries in order to qualitatively and quantitatively fill the gap between employment demand and supply.

The Reform of the Higher Education system launched by the Ministry of Education, University and Research has introduced new rules for the organisation of courses, consisting mainly in the rearticulation of university degrees in three year courses (with the possibility to further specialise in the subject of study by attending two additional years) and in the introduction of training credits, which quantify the number of hours needed by students to prepare an examination and imply a deeper commitment from teachers in the choice of learning materials and in the design of courses. Furthermore, the reform has increased the autonomy of universities in the design of the courses (although the general guidelines provided by the Ministry still have to be respected) and in the management of resources.

As far as the use of ICT in Higher Education is concerned, in the Action Plan for Human Capital[2] the Government envisages public-private sector coordination for research into ICT, promoting synergies among universities, private bodies and firms active in the field of ICT. This intention, declared in the Legislative Decree 297/99, translates into an overall investment of about 30 million Euro between 2001 and 2002 (of which 15 million Euro financed by the state) for:
– the design and implementation of university courses in economics and IT;
– the exchange of university researchers with industry and the development of academic spin-offs;
– the promotion of Centres of Excellence;
– the support of research in ICT.

The recently published document 'The Government guidelines for the development of the Information Society'[3], reinforces these aims through the identification of a set of action lines aimed mainly at upgrading the technological and networking infrastructure, fostering the link between universities and the labour world and establishing a national research programme for open source software.

The initiatives undertaken by the Government on one side and by universities on the other witness the increasing importance of Higher Education, which nowadays plays a threefold role in the Italian context: the traditional role of universities as education providers and cradles of research are now integrated with the role of bridging the educational needs of the students with the competencies required by the labour world. In order to fight unemployment it is first of all necessary to provide young people with the competencies required by the Knowledge and Digital Economy. For this reason, the Government and the Ministry of Education, University and Research in the first place are promoting the establishment of closer relations between industries, research bodies and universities and the creation of 10 Centres of Excellence devoted to ICT by 2002.

Many projects have already been developed, involving a number of Italian universities and research centres, and implying a close collaboration with industries. This has engendered the creation of networks and consortia involving representatives of research (namely, universities and research centres) and of industry with the aims to: transfer the innovative results of research to industry, provide consultancy services to companies, jointly develop innovation projects; foster the creation and development of new innovative firms in the ICT sector.

221

The increasing cooperation between research and industry is slowly reducing the gap between education and training, paving the way for a coordinated action among education and training institutions in order to fight unemployment and the ICT skills gap. It must be remarked, however, that a national policy for ICT in higher education has almost never existed in Italy: the only possible exception dates back to the early 1990s, when the use of ICT was supported in combination with the introduction of the three-year diploma in the Italian academic system. To this period dates back the creation of Nettuno as a national structure to support the individual universities, effort in this direction. After that period no national initiative has been launched; European programmes and the Association of Universities has sustained the interest of many Italian academic staff and university management teams in this domain, but, not by chance, very few Italian universities may be said to have something similar to an ICT policy.

Nevertheless, innovative and successful practice is not infrequent: the Consorzio Nettuno constitutes a significant gathering environment and the level of attention to e-Learning is growing in the political community, so it is likely that an e-Learning Plan will soon appear also in Italy and contribute to the consolidation of existing experiences.

13.3 Change Factors and University Motivations to Use ICT for Teaching and Learning

Among the most important trends affecting education and training identified by the Hectic project[4], the following are certainly relevant for Italian universities:

1   *Growing and more articulated demand for education and training,* which is not just increase of young people's demand (negative demographic trends are balanced by higher rates of participation in secondary and higher education), but rather increased demand by adults and social segments that were previously not interested or could not afford higher education and continuing training.

2   As a consequence of the first trend, *different expectations, motivations, skills and learning models, corresponding to the new classes and generations of learners,* have to be 'digested' by education and training institutions, which need to revise their teaching models, also through the use of ICTs, in the directions of increased autonomy of learners, different approaches to quality, differentiation of contents and learning services: in other words, growing levels of responsiveness to demand.

3   The revision of teaching models in the direction of autonomy of learners paves the way *for regular students to incorporate courses from other providers than their own university* in their degree programmes, and the competitive offer of full diploma curricula by foreign public and private universities or organisations, at the moment especially in the USA and Australia. This development also gives rise to discussions as to whether education is entirely a public good or should be regarded at least in part as a commodity to be consumed[5].

4   The increased demand of lifelong learning is also attracting *new providers* into education and training, many of which are 'for profit' organisations.

The increased importance of ICTs is further accelerating this process and contributing to accessibility of education supply coming from all over the world. Such tremendously increased competition on the education and training market is introducing new logics of investment in public institutions, generating new partnerships, among similar institutions in different geographic areas or among different organisations, to provide at the same time broader and more specialised learning contents and services.

5   Beside research and teaching, universities are more and more frequently required to expand their role in *accompanying policies for economic development and social inclusion* at regional, national and international level: here again the capacity to cope with new functions, new partners, new challenges is under discussion and the role of ICTs to support these innovations has to be taken into account as several institutions already do. In particular, ICTs play a fundamental role in the observed trend towards integration of formal and informal learning within a lifelong learning approach, so making university content and learning support services available to non-conventional target groups.

6   Among other consequences of both globalisation and the lifelong learning agenda the modified *role of certification* has to be considered. From one side it is more and more important that learners see their learning efforts and results recognised in a non-parochial context (inter-institutional, international): from the other side universities are certainly not the only organisations legitimised to provide titles and certificates: regional authorities, professional associations, training bodies, 'corporate universities', hardware and software providers, private training organisations are more and more active in the 'certification business', so threatening one of the perceived pillars of university survival in the long term.

The concrete reasons that move universities to pay attention and start to use ICT are different:

–   to meet the expectations of young students already familiar with ICT, and their future employers;
–   to experiment with the use of technology per se or as a way to change pedagogy towards a more student centred approach;
–   to facilitate access to courses by disadvantaged groups of students;
–   to address in a flexible way the public of adults that can pay for continuing education and so generate and consolidate a new line of academic activities and the related income;

223

— to access public funding occasionally available to support ICT use and flexible learning, especially in continuing training.

European rather than national policies have pushed many Italian universities towards inter-institutional cooperation in the domain of ICT and distance learning, and also encouraged collaboration with industry, social partners and regional authorities to establish a more responsive, demand driven, approach of academia to the needs of regional labour markets and development prospects.

13.4 Main Experiences of ICT Use in Italian Higher Education

The most significant and representative example of universities' commitment in innovation of the Higher Education system through the use of ODL and ICT is the *Consorzio Nettuno*, a network of Universities, companies and associations promoted by MIUR (Ministry of Education, University and Research) which is for the moment the first and only 'telematic and television' university providing courses and educational activities via the Internet and two dedicated satellite TV channels. Today, 38 Italian universities are members of the Consortium, together with 28 technological centres. With the support of 3,000 professors, 285 courses are provided via satellite video (thanks to the partnership with Raisat), with a total of 16,000 hours of video-lessons and 12,000 hours of interactive exercises provided through the Internet. The services are currently addressed to a total of 10,000 students, who can count on the support of on-line tutors for each of the subjects studied. Among the activities of the Consortium the initiative of the Mediterranean University should be mentioned, as it promotes collaboration and exchange of experiences among distance training universities of Mediterranean countries.

In the last two years, Italian universities have started launching courses on line, following the example of the Nettuno Consortium.

In 2000, the *Politecnico di Milano* launched the first on-line degree on IT engineering ('Laurea on-line initiative') with the collaboration of Somedia, specialising in the field of ODL, METID (innovative technologies and methodologies for learning) and OSSCOM (Observatory on communication of the University of Sacro Cuore, Milan). Due to the success of the initiative, the on-line degree has been re-launched this year.

The *University of Palermo* has recently launched an on-line degree on 'Technologies and Sciences for the Environment and Tourism' in cooperation with the Institute of educational technologies of the National Council of Research (CNR).

The *University of Torino* together with For.com (Formazione per la Comunicazione) provides an on-line degree in communication sciences, and many other universities are increasingly investing in ODL and e-Learning for the provision of University courses.

The *Università della Tuscia* has been the first Italian athenaeum launching the use of e-books for its students as learning tools. E-books published on line have both didactical and research content; e-book, available on-line on the University web site, allows downloading of materials, books, and conferences proceedings.

ICON *(Italian Culture on the Net)* is a Consortium gathering 23 Italian universities, created in 1999 with the support of the Italian Government and of the Ministry of University and Research in order to promote and diffuse the Italian language and culture in the world through on-line facilities. Through its didactic portal[6] ICON provides: university courses on the Italian language and culture (addressed to non-Italian students); Italian language courses; digital library, museum and encyclopaedias and interactive learning services.

The *University of Milan* enables access to information and short introductory on-line courses for first year students through its portal[7] and the *'Bocconi University'* has recently launched the B_Learning project, addressed to e-Learning professionals and making use of an e-Learning platform to support traditional didactic activities.

The *University of Padova* is facing the challenges of the Knowledge and Information Society with a deep process of innovation implying not only the integration of new technologies into didactics, but also the introduction of new courses and subjects aimed at providing students with all the necessary technical and behavioural skills requested by the labour market: in the academic year 2001/2002 a total of 94 degree courses completely renovated in their aims and didactic modalities were launched, and this year 60 Master programmes are offered in collaboration with industrial companies and bodies. In particular, such companies and bodies contribute to the design of curricula through their expertise and to the implementation of courses through their financial support.

The *Italian Conference of University Rectors (CRUI)* is deeply involved in the development of projects and initiatives aimed at fostering and facilitating the process of innovation and ICT integration in higher education: in the year 2001 the *Campus one*[8] pilot project was launched. This three-year project (2001-2004) addresses in particular the new triennial degree courses with the aim of supporting the reform of the Italian higher education system through technological and didactic innovation. All Italian universities (with particular reference to Southern Italy) are the target group of the project. The CRUI, together with Confindustria, MIUR (Ministry of University and Research), Regions, Trade Unions, Chambers of Commerce and CNEL (National Council of Economy and Labour) provide support to universities in the implementation of the reform with reference to: the national higher education system, the single university systems and the different degree courses.

Beside the increasing offer of on-line degree and Master courses, the emerging phenomenon of corporate universities should not be forgotten. As companies are becoming fully aware of the key role of human capital in their strategic development, the investment in corporate training initiatives is increasing, and big companies are starting to build their own training centres, where courses are designed to provide workers with the specific skills requested by the company or by external 'customer companies'.

Isvor Fiat[9], a service centre of the Fiat Group that operates in the fields linked to the improvement of human resources is one of the most important full-liner training companies in Europe, providing: off-the-rack training, tailor-made projects, training engineering (creation of structured training models), new technologies and new media for training and consulting for integrated management of training processes. Its customers are: small and large, public and private, commercial and industrial companies, but also local authorities and government bodies, international organisations, training bodies, foundations and universities.

Intesa Formazione[10], a training company belonging to the Intesa Banking group, created in 1999 provides, through the 'Intesa C@mpus' initiative, training courses targeted to the financial sector (employees and management), focussed on the acquisition of technical and core skills.

Scuola Superiore Guglielmo Reis Romoli[11] (Telecom Group), provides training courses targeted to human resources operating in the Communication Technology and Management sector.

The examples provided above show how universities and companies are not only investing in the innovation of learning provision, but also in the innovation of the subjects and contents of learning: both in graduate and postgraduate studies attention is now devoted to the development of technological and behavioural competencies aimed at training individuals to become flexible and able to face the fast changes of economy and society.

13.5 Major Trends and Future Scenarios

In order to envisage the future of the Higher Education system in Italy and more generally in Europe, it is necessary to consider the aspects of evolution of the whole education and training system. The L-CHANGE Study conducted in the framework of the IST project 'European Observatory on IST related change in learning systems'[12] has identified the following drivers of change, leading to a set of three possible scenarios as further described in this section.

In terms of primary 'change drivers' affecting education and training we can recognise two major forces[13]:

1   A push towards de-institutionalisation and 'marketisation' of education and training, according to which an increased autonomy of learners to choose and buy among a vast plurality of learning opportunities is not supported and facilitated very substantially by the conventional education and training organisations.

The identification of e-Learning as a likely area for quick market development is attracting more and more investors and major existing companies into the education and training areas, and also many newer and older universities are moving rapidly in this direction. The plurality of supply is progressively breaking a substantial public monopoly and is bringing strong price differentiation together with a 'consumer oriented' quality approach also affecting, to a certain extent, conventional institutions which have introduced learners' evaluation of courses, teachers and other components of the learning experience proposed.

2   The second main direction of change is related to innovation processes taking place within or by initiative of education and training: the lifelong learning philosophy, including a higher degree of integration among the different subsystems of education and training, the autonomy of learners and the shift from teaching-based to learning-based approaches also thanks to ICTs, the introduction of flexibility and quality elements based on a higher level of responsiveness to changes and needs of economy and society.

This 'driving force' has been expressing itself over a longer period than the first one, but with uncertain speed and results, until very recent years. Then, in particular the Internet has attracted the attention of many teachers and trainers as a far more interesting element to be introduced as an additional resource rather than stand alone multi-media products. The latter were more properly perceived as alternatives, which threatened teachers as direct providers of content.

In addition to these two driving forces of change, that have in common, among other things, an increased role of individual learners, we need to consider the tremendous inertia strength of education and training systems, which have very frequently absorbed some technological innovations in the past without substantially changing their way of working, so inhibiting both previously mentioned driving forces of change (see Figure 13.1).

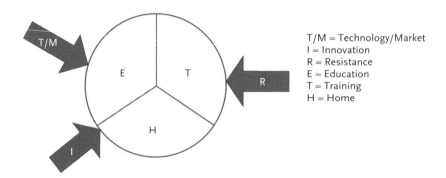

Figure 13.1 Driving forces of change

According to many observers[14], and also according to the first results of the analyses conducted within the L-CHANGE project, three main scenarios for future education and training can be envisaged, as a starting point to identify the extreme situations that can be expected, according to the respective full deployment of each of the two driving forces of change or the inertia force of education and training systems.

13.5.1 Scenario 1 – de-institutionalisation

In this scenario, conventional education and training institutions are losing importance to the benefit of new market participants who can assemble and offer a wide choice of contents and services to different segments of learners and consumers, with possible improvement of efficiency in the learning production process and improvement of learning options in terms of market relevance and formal 'aesthetic' quality, though not so secure in terms of content and pedagogical approaches.

In this scenario, the key assets are large availability of learning contents and capacity to distribute effectively to the highest number of potential learners and buyers, at home or at the workplace learning centres. The development of this scenario may be the result of policies supportive of privatisation of education and training or no policy at all, as market forces are working in this direction.

13.5.2 Scenario 2 – lifelong learning in place

In this scenario, innovation and transformation of education and training systems can effectively take place and lead education and training institutions to leadership of innovation through an increased provision of services that help learners to become autonomous and open-minded in the choice of their learning paths and personal development processes. The learning context and its capacity to generate communication added value among learners, between teachers, tutors and learners are the key quality factors of the learning experience. The service, rather than product, nature of learning is well perceived by individuals and 'organised' collective users, while content availability is not the key asset to be successful for learning providers: individual counselling, guidance, monitoring, evaluation, certification, community building and animation are the 'proximity services' that constitute a significant part of the quality of the learning experience.

Most e-Learning, in this scenario, is integrated in classroom or community or workplace learning, so it builds on existing groups of learners interested in sharing at least a part of their learning experience.

This scenario is the explicit objective of most European public policies focussed on innovation of education and training systems through the use of ICTs and on lifelong learning. Industrial participants are involved in the implementation of these policies, but mainly as partners of education and training institutions and within defined education and training policy aims. Behind these policies the principle of considering education and training as goods of public interest ('public good') that deserve public investment and cannot be left to market forces alone can easily be found.

An awareness of the risk of new social exclusion deriving from unequal distribution of access to ICT and learning opportunities is also inspiring lifelong learning policy orientations that are supportive to the establishment of this scenario.

13.5.3 Scenario 3 – inertia

In this scenario, the resistance of education and training systems to change in any of the directions towards which innovation programmes and market forces would tend to push is the characterising element.

Of course, if conventional institutions will not change substantially (they may well buy and use ICT, but simply continue, in a different way, the same ways of teaching based on content transmission) this does not mean that nothing will change around them in education and training. In fact a 'dual market' may develop in which title-oriented learners will continue to address conventional institutions while most competence-oriented learners will find other suppliers that can better serve their requirements or desires.

This scenario can emerge both as a result of 'protectionist' public policies which prevent private and generally new participants from interfering in formal provision of education and training (closed accreditation systems, emphasis on formal titles to access public administration, etc.) or as a failure in the implementation of innovation policies presented in scenario 2.

The suspicion that mainstream education and training programmes are actually contributing to dismantle what is a substantially good system is widely diffused among teachers, trainers and their representatives. Ideas of the need for cultural absorption into the particular professions and disciplines, or simply a paternalistic view of learners as everlasting children needing protection, pre-established structure and continuous support in order to be able to learn, are often behind resistance to innovation.

If we concentrate on Higher Education only, we can see a combination of the three extreme scenarios as highly possible, but affecting different parts of higher education provision:

- classic university paths leading to first and possibly second level academic titles are likely to resist innovation for a considerable number of years and defend their market position through their ability to deliver official titles and their direct access to public funding at national or regional level;
- third cycle and continuing education are more likely to introduce innovative approaches to learning provision, in which autonomous learners are accessing high quality contents and services, but also build 'knowledge workers communities' in which collaboration, exchange of experience, collective building of new knowledge are more important than content acquisition: the creation of a stimulating learning context becomes a more strategic asset for universities than the availability of attractive contents;
- the risk of 'consumerisation' is relatively lower in higher education than in other sectors of the education and training markets; however it may well emerge from a trivialisation of e-Learning approaches in the direction of access to 'top brand' contents. Some signs of this trend have been observed in recent years especially in the US market, but the real success of these ventures has still to be proved in most cases, and the recent MIT[16] move to make its 'contents' available free of charge on the Internet has certainly de-dramatised every claim of 'content is king', that was so popular in the 'booming' years of the new economy.

As a general conclusion we can expect a broader and better defined typology of higher education institutions to emerge in future, and even differentiated strategies to emerge in the same university to address different market requirements.

References

Aceto, S., Dondi, C. (2001). L-change Project, 'European Observatory on IST-related change in learning systems ' – *Country Report Italy*. (December).

Aceto, S., Barchechath, E., Dondi, C., *et al*. (2001). L-change Project, European Observatory on ist-related change in learning systems – *Scenario Design Report*. (July)

Floor, P., *et al*. (2002). *European Union Policies and strategic change for eLearning in Universities*. Hectic project. Coimbra Group.

Italian Minister for Innovation and Technologies. (2002). *The Government guidelines for the development of the Information Society*. (June).

OECD (2001). *Education Policy Analysis 2001*

Notes

1   Microsoft Observatory promoted by Microsoft in collaboration with Netconsulting and with the support of the Italian Ministry of Welfare, http://www.microsoft.com/italy/stampa/articolo_sez39info757.htm

2   http://www.palazzochigi.it/fsi/doc_piano_eng/cap2_eng.htm, par 2.7

3   Minister for Innovation and Technologies, June 2002

4   P. Floor *et al.*, Hectic project European Union Policies and strategic change for eLearning in Universities, Coimbra Group, 2002.

5   Recently the Association of Universities and Colleges of Canada (AUCC), the American Council on Education (ACE), the European University Association (EUA) and the Council for Higher Education Accreditation (CHEA) discussed this very important trend and signed a Joint Declaration on 'Higher Education and the General Agreement on Trade in Services'. See www.aucc.ca/en/international/bulletins/declaration.pdf

6   http://www.italicon.it

7   http://www.unimi.it

8   The 'Campus One' project (http://www.campusone.it) is financed by the Italian Government through the UMTS funds.

9   Isvor Fiat: http://www.isvor.it

10 Intesa Formazione: http://www.intesaformazione.it

11 Scuola Superiore Guglielmo Reis Romoli: http://www.ssgrr.it

12 ist-2000-26226

13 S. Aceto, E. Barchechath, C. Dondi et alii, Scenario Design Report, L-CHANGE Project – IST Programme, July 2001.

14 OECD, Education Policy Analysis 2001

15 2000 courses made of conferences in video format, notes, assignment and exams. Most of the 940 MIT professors are supporting the project which budget amounts to $100 millions. MIT President Charles Vest, explains that nothing could replace the richness of the campus life and is expecting amazing effects of his initiative. Source New-York Times 04/04/2001 and Wall Street Journal 05/04/2001.

14 Survey of ICT in Swiss Higher Education

Bernard Levrat[1], University of Geneva, Switzerland

14.1 Introduction

Switzerland is a small country with a tradition of local decision making in education at all levels. The central government in Bern funds directly two Federal Institutes of Technology in Lausanne and Zurich. Ten universities[2] depend on their home cantons but also need federal help to meet contemporary standards of research and teaching. This help comes in several forms: contributions to operating expenses and new construction, support of research projects by the Foundation for Scientific Research, and a number of incentive programs called special measures which encourage new developments within the universities.

In the past few years, a group of Engineering, Arts and Business Administration Schools has been organised into a network of Universities of Applied Sciences (UASS) which reports to the State Department of Economy while the older institutions depend on the Department of Interior (Health, Education and Scientific Research).

By tradition, all the above institutions have developed their teaching and their research to a large degree independently of each other. Recently, growing requests for contributions from the federal government have called for more coordination. All initiatives which promote such coordination have been welcomed by the various institutions.

14.2 Special Measures for Informatics

In the mid-1980s, the Federal Office for Education and Science convinced Parliament that a serious increase in support was needed to help universities provide their departments of computer science with modern equipment and to build a good network infrastructure linking all the higher education establishments.

The SWITCH foundation was established in 1987 by the Swiss Confederation and the (then) eight university cantons both to promote modern methods of data transmission and to establish and to operate an academic and research network in Switzerland. The SWITCH head office, located in Zurich, operates some 20 sites throughout Switzerland as well as the central SWITCH system at the ETH Zurich Computer Center. Today, the UASS are connected to SWITCHlan via their own network which, for reasons of readability, is not represented on the map provided by SWITCH at http://www.switch.ch/network/national.html.

SWITCH is a member of TERENA (Trans-European Research and Education Networking Association) and of the TEN-34 Consortium which has set up a pan-European 34 Mbit/s network. SWITCH is a shareholder of DANTE.

Complementing these inter-institutional network developments, most of the Higher Education establishments have built their own local area networks and acquired a substantial number of workstations for student use in all disciplines.

14.3 Use of ICT in Education: the Pioneers[3]

The potential of computers in such an infrastructure encouraged increased expectations about improving education with technology. Some institutions tried to look into other aspects of the introduction of computers in education: psychology, pedagogy and social impact. Occasionally, the concept of 'evaluation' can also be found.

14.3.1 The Federal Institutes of Technology (FITs)

Early on, convinced that synchronous communication has an important part to play in education, the two FITs, situated in Lausanne and Zurich, joined forces to develop TELEPOLY[4] which became operational in 1996, and extended to Basel in 1998. It uses advanced technology to provide 'live' transmission of lectures with high quality audio and video and 'screen sharing' tools for electronic teaching aids in a distributed environment.

The research and development project CLASSROOM 2000[5] was started in November 1997 and provides the structural, technological, and didactical framework for the implementation of new learning technologies in the form of modular courses for engineers and technicians. The project runs under the scientific management of the Computer Science Department (DI-LITH) at the Swiss Federal Institute of Technology in Lausanne (EPFL), is coordinated by NDIT/FPIT[6] and realised together with a consortium of Swiss universities, universities of applied sciences and companies.

14.3.2 NET: a center of ETHZ to disseminate new learning and teaching technologies

To promote the integration of new information technology in teaching, the Network for Educational Technology (NET) arose as an initiative from the Center for Continuing Education and the Didaktikzentrum in 1996. It provides information, consulting, and support to instructors. It also initiates and supports projects. The structure has changed and NET has received a permanent status within ETHZ. NET was also the only provider of consulting and support for instructors of the University of Zurich until 1999, when the ICT Fachstelle was founded within that university.

14.3.3 ARIADNE: a European project involving EPFL and Lausanne University

235

ARIADNE[7] is a research and technology development project pertaining to the 'Telematics for Education and Training' sector of the 4th Framework Programme for Research and Development (R&D) of the European Union. The project focuses on the development of tools and methodologies for producing, managing, and reusing computer-based pedagogical elements and telematics-supported training curricula. Since 1996, with the support of the Swiss Federal Office for Education and Science (OFES), both EPFL (the lead institutions) and the University of Lausanne have been very active in this ambitious R&D project (approximately 100 man/years invested in two phases of the EU Project).

In March 2000, the ARIADNE Foundation was created to exploit and to continue developing the results of the ARIADNE project. Validation of the tools and concepts took place in various academic and corporate sites across Europe and was encouraging enough to go ahead with this idea of non-commercial exploitation. The ARIADNE research group is also very active in the standardisation community, in particular with regard to the so-called 'Standard for Learning Object Metadata' recently adopted by the IEEE.

14.3.4 'Centre NTE[8]' (New Technologies and Teaching) at the University of Fribourg

Created in 1996 by the Rectorate of the University of Fribourg, the Centre's mandate is to develop the use of the NTIC in university teaching and learning and to observe its impact
for both teachers and students. It helps teachers develop learning material promoting student autonomy and active participation. To gain experience, the Centre has contributed to several pilot experiments that have been evaluated.

14.3.5 TECFA[9] at the University of Geneva

Created in 1989 by the Faculté de Psychologie et des Sciences de l'Education of the University of Geneva, TECFA is a research and teaching unit active in the field of educational technology. TECFA's research covers a large area of interests including the following: cognitive issues in learning technology, computer-supported collaborative learning, virtual learning environments, computer-mediated communication, information systems in education, and distance education. TECFA has supported many developments and participated in several EU projects. Since 1994, a postgraduate diploma (DESS) in educational technology has been offered (under the acronym STAF for 'Sciences et Technologies de l'Apprentissage et de la Formation') which combines face-to-face activities (6 weeks per year) and web-based activities. TECFA has been a pioneer in designing a virtual campus, its own version promoting a constructivist approach.

14.4 Special Measures for ICT in Education

14.4.1 Motivation

In 1998, a proposal for incentive measures from the central government to promote the development of Web-based material was accepted by decision-makers and teachers alike. They worried about the changes that global access to Internet resources will bring to education. They were aware that small-scale experiments do not scale up easily. Many teachers would like to take advantage of the new information and communication technologies; they are using e-mail, newsgroups, Web pages, and popular software to modernise their communications with their students, but some were also interested in participating in larger schemes to produce material that would be competitive with what is bound to appear on the market: they are the ones who responded to the calls for participation in the programme below.

14.4.2 *The Swiss virtual campus program*

In its message to Parliament concerning the funding of Higher Education for the period 2000-2003, the Swiss Government included a credit of SFr 30 million (20 million euros) for funding the Swiss Virtual Campus with the following objectives:
- Higher Education Networks to promote collaboration between institutions;
- Learner-Centric Education to entice teachers to explore a new pedagogical dimension;
- Competitiveness of Swiss Higher Education Institutions to encourage the development centres to become producers of high quality learning material.

The projects that are jointly funded by the universities and by this program, the FITs and the UASs are invited to contribute their own resources, have to abide by the principles of the 'Swiss Virtual Campus' which are summarised in the five following criteria:
- favour cooperation between Higher Education establishments (traditional Universities, Swiss Federal Institutes of Technology, and Universities of Applied Sciences);
- highlight clear pedagogical objectives;
- participate in the selection and use of common tools on stable platforms;
- present implementation plans with evaluation criteria;
- obtain, from the start, (1) financial support (matching funds) from the institutions interested in the project and (2) the commitment to integrate the use and maintenance of the project's 'deliverables' into the institutions' normal planning once the project's development is finished.

A public call for participation produced 155 letters of intent by September 1999. An international group of experts selected 55 of them to present a detailed proposal: 27 were accepted and started in July 2000. A second series produced 58 proposals with 22 accepted starting in July 2001. Each project involves at least three institutions (leader and partners) which pledged to let their students take the courses for credit after an appropriate evaluation.

The project management of the SVC is organised in the following way. The Swiss University Conference is in charge of the project management with the finances audited by the Federal Office of Education and Sciences. The main decisions concerning the running of the project are prepared by a Steering Committee chaired by Professor Peter Stucki from the University of Zurich.

To help the projects, mandates were given to cover technical, legal, pedagogical, and organisational issues. The Edutech[10] organisation maintains an evaluation of platforms, runs seminars for project managers, and gives whatever technical support is requested. In order to provide pedagogical support to the SVC projects, to make an inventory of the projects' pedagogical practices (insisting upon the exploitation of the innovative and interactive potential of ICT), and to set the bases of an evaluation framework that would permit assessment of the innovative nature of e-Learning pedagogy, the two pedagogical mandates (InterSTICES[11], for the French speaking community and eQuality[12], for the German speaking part) visit the projects and organise different activities integrating accompaniment and instruction. InterSTICES early experience suggests that, beyond pedagogy per se, many factors need to be taken into consideration for each of the various participants (e.g. professors, development team, support team, learners, institutions, etc.): their representations, beliefs, attitudes, abilities, experience with both technology and with active pedagogy. Institutional and societal factors should be recognised as major factors of influence (programmes orientation, evaluation procedures, team/project management issues).

14.4.3 The digital libraries consortium

In parallel with the SVC, the Swiss authorities set up a programme to give access to the best selection of products at the best possible price, to set up national licences for the use of information resources in Swiss higher education institutions, to develop a project structure common to all partners, and to organise international cooperation with other consortia and product providers.

The total costs budgeted for 2000-2003 come to around 19 millions SFr. with a federal contribution of 7 millions from the central government. The detail of the many databases and electronic journals available can be found on the consortium homepage[13].

14.5 The Situation Today

All Swiss higher education institutions forecast an increased role for Information and Communication Technologies (ICT) in teaching and learning; the Federal Institutes of Technology and several universities started support centres for pedagogy, multi-media, and software production. These centres play a key role in implementing the Swiss Virtual Campus (SVC). Switzerland is small enough that we can present in an appendix a list of all the SVC projects including their title, leading institutions and, when they are available, the URLs of their homepages.

14.5.1 ETHZ

Despite the fact that no specially dedicated funding has been available, e-learning competency has spread all over the individual departments by the coordinating activity of NET, its consulting activity as well as its limited support to any interested faculty. Financial support comes now from two sources: Fonds Filep (about 5 million SFr/year) and from the FITS Council for specific projects.

– ETH World: a virtual space for everyone associated with ETH Zurich
  ETH World is a strategic initiative to prepare ETH Zurich for the information age. Its objective is to create a universal virtual communication and cooperation platform, supporting the activities of everyone working or studying at ETH. ETH World will augment the present physical locations, 'Zentrum' and 'Hönggerberg', by a virtual space, creating a third, virtual campus for ETH Zurich.
  The development of ETH World is a project running until 2005. In 2000 an international conceptual competition was organised to seek ideas and a master plan for the implementation of ETH World. Work is now underway to implement the results of the competition. Parallel to this, ETH World is being built through a growing number of individual projects, developing e-learning, research tools, information management, infrastructure elements, and community building.
– SVC Projects:
  The Department of Seismology and Geodynamics is co-leader of 'Dealing with natural hazards' and ETH is partner in 12 projects.

239

14.5.2 EPFL

'EPFL on line'

In June 1999, the EPFL adopted a plan of action concerning the new Information and Communication Technologies for teaching. Called 'EPFL on line', this institutional programme aims at building a range of flexible and distance teaching, gradually offering a series of educational modules. Thirteen priority projects are involved and a new R&D centre for learning technologies will be created by the end of 2002.

SVC Projects

Several priority projects supported by the FITS Council, EPFL, and the federal Swiss Virtual Campus programme are being carried out and experimented with. EPFL is leader in 'i-Structures: Interactive Structural Analysis by Graphical Methods', co-leader in 'Dealing with natural hazards,' and partner in 4 projects.

14.5.3 University of Basel

Very active in the promotion of e-learning, the University of Basel has established a Network for Learning Technologies and E-Learning (LearnTechNet) for ICT use in higher education. The LearnTechNet consists of different units within the university. The Department of Teaching, which is part of the Vice President's Office for Teaching, offers support for didactics, evaluation and integration of on-line modules into the curricula; the Computing Center of the university supports the projects in all technical concerns, and the New Media Center provides support and services for development and design of multi-media.

The University of Basel leads 6 projects of the SVC: 'The Virtual Nanoscience Laboratory (Nano-World)', 'Latinum Electronicum', 'BOMS – Basics of Medical Statistics', 'Financial Markets', 'TropEduWeb: Public and International Health and Epidemiology with special reference to Tropical Medicine', 'Course of Pharmaceutical Chemistry in a Virtual Laboratory' and is partner institution for a further 14 SVC projects.

14.5.4 *University of Bern*

The University of Bern's central IT-facility hosts a WebCT server and offers support for its use. Bern leads four SVC projects: 'VITELS – Virtual Internet TElecommunications Laboratory', 'artcampus – introduction to the history of art', 'ViLoLA – a Virtual Logic Laboratory', 'OPESS – Operations Management, ERP- and SCM- Systems'. Additionally, it is a partner in another 19 SVC projects.

14.5.5 *University of Fribourg*

Fribourg's 'Center for New Technologies and Teaching' is very active supporting local and SVC projects. It was selected to establish and to run the www.edutech.ch resource center and to maintain the www.virtualcampus.ch official site. Fribourg leads three SVC projects: 'European Law On-line', 'A Web-Based Training in Medical Embryology', 'Antiquit@s – Ancient history learning project' and is partner in another 18.

14.5.6 *University of Geneva*

Leader of two projects ('Computers for Health' and 'SUPPREM: Sustainability and Public or Private Management') and partner in 13 other projects of the SVC, Geneva is quite active in developing new forms of applications of ICT to education. Its Rectorate has appointed a coordinator, methodological support is offered by TECFA and technical expertise can be found within its Computing Services Division.

Several courses and seminars are offered as synchronous distance courses using a set of tools for transmitting voice, documents, and computer screens. Intended at first for auditors in the local community, these tools have proved their value for delivering lectures by members of the Faculty of Medicine to developing countries. This form of e-learning may solve some of the problems related to tutoring distance learners. An example of such an effort is the 'La Francophonie' funded project www.universante.org, in which students and tutors from four different countries (Lebanon, Tunisia, Cameroon and Switzerland) work together on public health problems in an on-line collaborative learning environment.

241

14.5.7 *University of Lausanne*

In addition to being a partner in 11 SVC projects, Lausanne is project leader for 9: 'General chemistry for students enrolled in a life sciences curriculum' 'SOMIT – Sport Organisation Management Interactive Teaching', 'Objective Earth, a planet to Discover', 'eBioMed – Biomedical sciences teaching modules', 'Information Theory', 'Immunology on-line: Basic and Clinical Immunology', 'Marketing On-line' , 'VSL: Virtual Skills-Lab', 'E-CID: An on-line laboratory for Spanish grammar learning'. This activity started well before the SVC programme and a sisable amount of on-line learning can be found in the MBA program as well as in other disciplines of the HEC Faculty. Lausanne has set up a support centre (CENTEF[14]) with over 20 full-time staff.

14.5.8 *University of Luzern*

Leader of the SVC project 'Introduction to Systems Theory and Analysis for the Social Sciences' and partner in three, Luzern is just starting to introduce ICT in its programmes.

14.5.9 *University of Neuchâtel*

While the introduction of ICT in teaching and learning was not a high priority for the Rectorate, it has agreed to co-finance the project called 'Do it your soil' and nine projects where it is a partner.

14.5.10 *University of St. Gallen*

Leading the SVC project 'Family Law On-line' and partner in three other SVC projects, St. Gallen is an advanced user of ICT in its business and finance courses.

14.5.11 *University of Svizzera Italiana*

The University was created in 1996. Its Faculty of communication sciences is developing expertise in the use of new media in education combining technical expertise with more pedagogically oriented reflection (e.g. through its new media laboratory). USI leads a doctoral school on the use of new media in education, financed by the Swiss National Science Foundation in cooperation with the universities of St. Gallen, Geneva, Fribourg, and Neuchâtel. USI leads the project 'Swissling: – A Swiss network of Linguistics Courseware' and is partner in 5 projects. It also received a SVC mandate: 'Educational Management in the SVC'.

14.5.12 University of Zurich

Founded in 1999, The University Center for Learning Technologies[15] (ICT Fachstelle), part of the Vice President's Office for Teaching, initialises, organises, and supports e-learning- innovations at the University of Zurich. Besides SVC projects, the center supports 87 ICT projects of the University. The University's e-learning funds for the years 2000-2004 amounts to SFr. 24 million. At the center, a staff of 8 people manages a pool of 70 full-time jobs for people engaged in project developments.

The University of Zurich is leader in eight SVC projects: 'DOIT – Dermatology on-line with interactive technology', 'Corporate Finance', 'Methodological Education for the Social Sciences', 'Basic course in Medicine and Pharmacology', 'ALPECOLe: Alpine ecology and environments', 'GITTA: Geographic Information Technology Training Alliance', 'Basic and Clinical Pharmacology: A Platform for Students in Medicine and Pharmacy', 'A comprehensive Internet course on Alzheimer's disease and related disorders for medical students' and partner in 14 other SVC projects.

243

14.5.13 The universities of applied sciences

The seven Universities of Applied Sciences admitted their first students only four years ago. They were created from a networking of existing institutions which underwent considerable changes to fulfil the ambitious goals assigned to them. They financed their projects with their own resources distinct from the budget under the control of the Steering Committee.

The SVC gave them the opportunity to consolidate some existing cooperation as well as to develop new material. 11 projects are under UAS leadership 'e-Ducation in environmental management', 'POLE – Project Oriented Learning Environment', 'Postgraduate Courses in a Hybrid Classroom using Mobile Communication', 'eduswiss on-line (EDOL)', ' Modeling and Simulation of

Dynamic Systems – A collection of applied examples ',’Forum New Learning', 'MACS: continuous education modules', 'Internet based course on Fundamentals of Signals and Systems', 'Development of a module entitled H bridge from the power electronics syllabus', 'Design of a CAL package teaching students effective information retrieval strategies', 'Basic Principles of Oecotrophology / Home Economics and Nutrition', 'Development of a Web based course for the application of the finite element analysis (FEA) in structure mechanics'. They are partners in 8 other SVC projects where the leader is not a UAS.

An extension to their participation to the SVC is a BBT project, the Creatools[16] which was funded earlier this year to support e-learning projects exclusively of Universities of Applied Sciences. 22 projects were selected at the beginning of March. Unlike the projects of the first two phases of the Swiss Virtual Campus, these projects do not require cooperation between different universities. Instead these must be projects in which at least one faculty member and two students take part.

14.6 Future Plans

A new proposal covering the period 2004-2007 is being prepared. It will have to be approved by Parliament and some changes may still occur. It asks for SFr 50 millions over 4 years and intends to achieve the following goals:
- provide maintenance for the successful projects from the current period and offer support to their users (9,120,000 SFr.);
- set up in each institution a professional production team (15,200,000 SFr.);
- fund new projects selected after another call for participation (22,800,000 SFr.);
- provide for coordination, mandates, and program management (7,580,000 SFr.).

In addition, it is foreseen that, with the help of SWITCH, a true authentication and authorization infrastructure will enable a large community of users to access seamlessly a wide range of services provided by the higher learning institutions. These institutions can contribute to lifelong learning by letting individuals access resources that will further their personal or professional development. Face-to-face as well as remote tutoring will have to be organised under terms yet to be defined.

The results of the pedagogical mandates will shed some light on the culture of innovative pedagogy in e-learning from multiple perspectives. Doing so, tightly interwoven discovery and instructional activities can lead to deep and sustainable changes in vision, abilities and practices, opening the way to a new e-learning culture and facilitating unavoidable societal changes.

Finally, good multi-lingual material produced within Switzerland should find its way into the world market, either with the support of commercial organisations that will work for profit or, with governmental help, to give individuals in less advanced countries access to learning resources which could contribute to bridge an ever-widening digital gap.

14.7 A Critical Assessment

The main merit of the programme is to raise the level of awareness of the Swiss Higher Education authorities from nice words to real budgets. All institutions have to provide matching funds and some of them have gone much further like, for instance, Zürich with almost 100 projects of its own.

Out of the 50 projects financed by the programme, a majority will reach the ambitious objective of producing full courses that students can take for credit on the Internet. But it has not happened yet and there are many unknowns in the future attitude of Faculty and students. The importance of tutoring has been recognised but its financing, particularly its sharing among institutions, has not been solved. The use of the SVC learning material for 'blended learning' is to be expected. It is hoped, however, that real distance learning as specifically designed in the programme, will also happen. It will certainly be the case in medicine where new paradigms for teaching and learning are developing around the concept of 'project based' learning.

245

The main worry is sustainability. The development costs cannot be sustained in a maintenance phase. Yet, the programming and pedagogical competencies that have been assembled around each project are necessary for their future adaptation and evolution. Some of the specialists will find a place in the newly created competence centres but many will move to other jobs outside universities.

Current uncertainties about future funding of projects must be overcome if the Higher Education establishment want to keep the highly skilled personnel that took them years to bring to such a level of specialisation. In my opinion, there is a need for new job definitions and profiles. People developing and maintaining quality e-learning materials should achieve recognition and career paths similar to librarians. Hopefully, it will entail the same professionalism and dedication.

Two years from now, the degree of success of the programme in reaching its stated objectives will be measurable. For the time being, the Steering Committee and the University Commission are trying to identify the main pitfalls and to avoid them.

Acknowledgments

I want to thank all the people who sent corrections or additions to the first draft of this paper as posted on the net. They are Gudrun Bachmann (Universität Basel), Gerald Collaud (University of Fribourg), Pierre Dillenbourg (TECFA), Martina Dittler (Universität Basel), Ronald Greber (Universität Bern), Bengt Kayser (University of Geneva), Chandra Holm (UAS Solothurn), Elaine.McMurray (FPFL), Benedetto Lepori (Università della Svizzera italiana), Eva Seiler Schiedt (Universität Zürich), Christian Sengstag (ETHZ), and Jacques Viens (TECFA) with a special mention to Jacques Monnard (Edutech) and to Cornelia Rizek-Pfister (coordinator of the Swiss Virtual Campus Program).
Stephen D. Franklin from UCI, agreed to improve the English of the paper and made some valuable suggestions on the contents.

Notes

1   http://cui.unige.ch/~levrat
2   Basel, Bern, Fribourg, Geneva, Lausanne, Luzern, Neuchâtel, Svizzera Italiana, St. Gallen, Zurich
3   It is difficult to search the Web for documents on this early phase because they have been replaced by more recent ones which do not mention the origins.
4   http://www.iam.unibe.ch/~rvs/events/nlt/talks/haenni.pdf
5   http://spp-ics.snf.ch/spp_ics/results/r051362_Classroom_sppics-proc2000.pdf
6   http://www.ndit.ch/5000/5500/d_5500.htm
7   http://ariadne.unil.ch/
8   http://nte.unifr.ch/presente.asp
9   http://tecfa.unige.ch/tecfa/tecfa-overview.html

10 http://www.edutech.ch
11   http://tecfa.unige.ch/proj/cvs
12 http://www.equality.unizh.ch/
13 http://lib.consortium.ch/Produkte_e.html
14 http://www.centef.ch/
15 http://www.ict.unizh.ch/index.en.html
16 http://www.creatools.ch

Appendix Chapter 14   The SVC projects

The tables below are extracted from the description of projects found on the Swiss Virtual Campus official home page www.virtualcampus.ch. The Universities of Applied Sciences are called Fachhochschule in German and Hautes Ecoles Spécialisées in French.

Arts and Humanities

Uni Fribourg	Antiquit@s – Ancient history learning project	http://nte.unifr.ch/cvs/antiquitas/
Uni Bern	artcampus	www.artcampus.ch/html/de/index.htm
Uni Lausanne	E-CID: An online laboratory for Spanish grammar learning	
Uni Luzern	Introduction to Systems Theory and Analysis for the Social Sciences	
Uni Basel	Latinum Electronicum	www.unibas.ch/latinum-electronicum/
Uni Zürich	Methodological Education for the Social Sciences	www.methpsy.unizh.ch/mesosworld/
Università della Svizzera italiana	SWISSLING – A Swiss network of Linguistics Courseware	www.swissling.ch/

Business Administration

Hochschule für Technik und Architektur , Biel	e-Ducation in environmental management	
Uni Lausanne Uni Basel	Marketing Online	
Zürcher Fachhochschule, Wädenswil	Basic Principles of Oecotrophology / Home Economics and Nutrition	http://svc.hswzfh.ch/
Uni Bern	OPESS: Operations Management, ERP- and SCM-Systems	http://opess.ie.iwi.unibe.ch/
Uni Lausanne	SOMIT: Sport Organisation Management Interactive Teaching	www.somit.ch/
Uni Genève	SUPPREM: Sustainability and Public or Private Environmental Management http://ecoluinfo.unige.ch/recherche/supprem/	

Economics, Finance and Law

Uni Zürich	Corporate Finance	www.getinvolved.unizh.ch/
Uni Fribourg	European Law Online	
Uni St. Gallen	Family Law Online	
Uni Basel		Financial Markets

Education

Fachhochschule Bern	eduswiss online (EDOL)	www.edol.ch/
Fachhochschule Bern	Forum New Learning	www.fnl.ch/
Scuola universitaria professionale della Svizzera italiana	MACS: continuous education modules	http://virtualcampus.supsi.ch/macs/
Fachhochschule Aargau	POLE – Project Oriented Learning Environment	

Engineering and Information Technology

Haute Ecole de Gestion, Genève	CALIS – Computer-Assisted Learning for Information Searching	www.geneve.ch/heg/rad/projets/campus_virtuel.html
Fachhochschule Aargau	Development of a web based course for the application of the finite element analysis (FEA) in structure mechanics	
Uni Zürich	GITTA: Geographic Information Technology Training Alliance	www.svc-gitta.unizh.ch/index.html
FHBB Nordwestschweiz	Development of a module entitled 'H bridge' from the power electronics syllabus	www.leistungselektronik.ch/
EPFL	i-Structures : Interactive Structural Analysis by Graphical Methods	http://i-structures.epfl.ch/
Zürcher Hochschule Winterthur	Internet based course on Fundamentals of Signals and Systems	
UAS Biel	Postgraduate Courses in a Hybrid Classroom using Mobile Communication	www.vcs.fhso.ch/
Uni Bern	Virtual Internet TElecommunications Laboratory of Switzerland (VITELS)	www.vitels.ch/

Environmental and Life Sciences

Uni Zürich	ALPECOLE: Alpine ecology and environments	www.geo.unizh.ch/virtualcampus/alpecole/
Uni Basel	Course of Pharmaceutical Chemistry in a Virtual Laboratory	
Institut für Schnee &Lawinen-forschung	Dealing with natural hazards	www.geo.unizh.ch/virtualcampus/nathaz/
Uni Neuchâtel	Do it your soil	www.unine.ch/bota/levp/teaching/campus.html
Uni Lausanne and EPFL	General chemistry for students enrolled in a life sciences curriculum	www.centef.ch/chimie/index.htm
Uni Lausanne	Objective Earth, a planet to Discover	

Medicine

Uni Zürich	A comprehensive internet course on Alzheimer's disease and related disorders for medical students	
Universitäts-spital Zürich	Basic and Clinical Pharmacology: A National Platform for Students in Medicine and Pharmacy	
Uni Zürich	Basic course in Medicine and Pharmacology	
Uni Basel	BOMS - Basics of Medical Statistics	www.boms.ch/
Uni Genève	Computers for Health	
Universitätsspital Zürich	DOIT - Dermatology online with interactive technology	
Uni Lausanne	eBioMed – Biomedical sciences teaching modules	
Uni Fribourg	A Web-Based Training in Medical Embryology	www.unifr.ch/histologie/svc/introduction.html
Uni Lausanne	Immunology online: Basic and Clinical Immunology	
Uni Basel	TropEduWeb: Public and Inter-national Health and Epidemiology with special reference to Tropical Medicine	
Uni Lausanne	VSL: Virtual Skills-Lab	

251

Physics, Mathematics and Informatics

Uni Lausanne	Information Theory	
Hochschule Wädenswil	Modelling and Simulation of Dynamic Systems – A collection of applied examples	
Uni Basel	The Virtual Nanoscience Laboratory ('Nano-World')	www.nano-world.org/
Uni Bern	ViLoLa: a Virtual Logic Laboratory	www.vilola.unibe.ch/

15 Competition, Collaboration and ICT: Challenges and Choices for Higher Education Institutions

Robin Middlehurst, University of Surrey, United Kingdom

15.1 Introduction

Several recent policy analyses of the higher education context in different parts of the world (CVCP, 2000; Collis and Gommers, 2001; Newman and Couturier, 2002) have focussed on the strategic responses of institutions to competition from new providers. These analyses have also noted the ways in which higher education institutions are using information and communications technologies (ICT) individually and cooperatively to deal with such competition. In practice, it is debatable whether institutional strategies for the application of ICT are a *direct* response to real or perceived competition from new providers. We shall address that question later by examining institutional rationales for their ICT applications. The policy analysts suggest that the emergence of new providers and new kinds of provision in higher education is part of wider economic and social changes. These changes are being fuelled both by the spread of information and communications technologies and the ever-increasing expectations of the benefits flowing from their use. Both new and existing providers are responding to the opportunities offered by the 'information revolution'.

This paper examines institutions' strategic responses to ICT and to new providers from a number of angles. The first section considers the macro- and micro-environments that affect institutions and that are likely to influence their ICT strategies. The second section includes an examination of national and regional responses to ICT applications in higher education before addressing institutions' strategic choices in the third section. For comparative purposes, the fourth section offers a brief look at company and competitors' strategies. The fifth section considers the implementation issues facing institutions as they seek to turn policy into practice, while the final section raises questions about the wider impact of ICT, not only on students and staff, but also on images of the 'higher education institution' as an organisation.

15.2 The Wider Context: Macro- and Micro-environments

My starting position is that universities and colleges are affected by and interact with their macro- and micro-environments so that institutional strategies and responses are not shaped in isolation of a wider context. In reality, institutions exist in an 'open system' of multiple interactions and influences. Higher education's macro-environments are global, regional and national and include political, social, economic and technological drivers and trends. Institutions' micro-environments are localised, reflecting the particular regulatory, funding and historical contexts of each institution. While there may be similarities in institutional applications of ICT, there will also be differences, arising from the real or perceived position of institutions in relation to their environments. Such differences, as we shall see later, are particularly marked between the developed and developing world. There will also be differences in national and institutional responses to new providers arising from these varied contexts.

Despite differences in labelling, there is wide agreement between authors of 'futures' studies' on the main drivers for change at a macro-environmental level (Tate, 2000) although much debate about the impact on countries, businesses and societies. From his analysis, Tate suggests that the major forces driving change are a combination of developments and trends that are technological, global, social, demographic and political (Tate, 2000: p.7). He argues that these drivers are already either fact or largely incontestable, open only to marginal influence at national levels (except, of course, in the event of unforeseen catastrophes). Tate lists the main forces driving change as developments in information and communication technologies and bio-technologies, the global environment and global economy, demographics, political developments and changes in peoples' attitudes and values as the wider forces of the 'post-modern information age' impact upon them. Several kinds of impact arise from these forces of change. By way of illustration, I have taken three of Tate's drivers to illustrate the types of impact that they have and their relevance for higher education.

Driver: Developments in information and communications technologies.
Type of Impact: Developments in ICT facilitate interconnectedness and convergence between organisations, sectors and nations as well as huge expansion in the creation of, and access to, codified knowledge.
Relevance to higher education: Both governments and individuals see higher education as central to the realisation of a 'knowledge-driven economy' and

developments in ICT make it easier for higher education to collaborate with a range of partners and connect with knowledge communities locally and globally.

Driver: Trends in the global economy.
Type of Impact: Growth in world trade is likely to continue along with an expansion of capital markets and flows. Liberalisation in trade barriers as well as increasing standardisation in business language, computing and telecommunications are occurring as part of these trends.
Relevance to higher education: The emergence and expansion of for-profit higher education and in levels of competition between providers is closely associated with global economic developments. Liberalisation and standardisation pressures that impact on businesses also impact on the delivery of higher education, particularly in trans-national and ICT-based contexts.

Driver: Demographics.
Type of Impact: Demographic changes point to continued growth of the world's population (to approximately 8 billion by 2020, according to United Nations predictions, with just over one billion in the industrialised nations) but with significant differences in the global balance between nations and regions. Italy will have more than 30% of its population over 50 by 2025 while the Philippines will have about 50% under the age of 40 (USCENSUSBUREAU, 2002).
Relevance to higher education: Demographic realities have implications for levels of demand for education, and for higher education, both in terms of initial studies and life-long learning, when linked to increases in per capita income (Olsen, 2002)

255

Higher Education analysts such as Cunningham and colleagues (2000) and Van Damme (2002) broadly agree with analyses of the drivers for change at the macro-environmental level. They also point to some specific forces that are affecting higher education. Newman and Couturier (2002) suggest that some of these forces show remarkable similarities across the globe. Four areas are highlighted by these authors: expanding enrolments; the growth of new competitors, virtual education and consortia; the global activity of many institutions; and the tendency for policy makers to use market forces as levers for change in higher education (Newman and Couturier, 2002: p.5).
Expanding enrolments are a global phenomenon (World Bank, 2000) although numbers vary across countries and regions. The World Bank reports that the number of tertiary students worldwide doubled in 20 years, growing from 40.3 million students in 1975 to 80.5 million students in 1995 (World

Bank, 2000: pp.107, 111). Other researchers have sought to extrapolate from current figures to predict future global demand. For example, Blight's IDP study (1995) forecast that global demand would grow from 48 million enrolments in 1990 to 159 million in 2025. Since then, these estimates have been revised upwards (Bohm & King, 1999). Blight's original study also showed wide variations between regions with demand from Africa modelled to grow from 2 million to 15 million enrolments (5.8% per annum) while Europe would only grow at 1.4% (including 0.7% for Western Europe). Bohm and King's reworking of Blight's data for China show much higher levels of demand for 50 million enrolments by 2020, with unmet demand predicted to be standing at 20 million by that date (Olsen, 2002).

Reports on 'Borderless Education' (CVCP, 2000; Cunningham et al. 1998, 2000) have charted the emergence of new competitors, virtual education and consortia as a response to the demand for higher education, the spread of information and communications technologies and the needs of knowledge economies. The 'new competitors' (some of which have been in existence since at least the 1970s, while others are of more recent origin) include private higher education institutions, for-profit institutions, corporate universities, media and publishing businesses, education brokers and consortia of various kinds. Private higher education institutions have existed in many countries (such as Japan or Thailand) for years, but in other countries they are either a new phenomenon or have grown significantly. For example, Poland had almost no private HE institutions in 1989 and now has 180 institutions enrolling one-third of all higher education students, while in Malaysia more than 400 private institutions were set up between 1992 and 1999 (Newman & Couturier, 2002 p.5).

The for-profit institutions have targeted and are capturing a new market, of working adults, with the University of Phoenix leading the field in the USA with more than 100,000 students. Corporate Universities focus on a similar client group in the form of their own employees as well as customers and suppliers, and with some open programmes. Today, there are more than 2000 corporate initiatives in the US alone with growth also in Europe (Taylor & Paton, 2002).

Media and publishing businesses as well as educational brokers are also a feature of the borderless education market-place. Many of these companies (for example, News Corporation, Pearsons and Thompson Learning) collaborate with higher education institutions where they wish to promote courses, programmes and other learning resources that can contribute or lead to awards. Consortia of institutions, usually involving corporate partners, are also a growing phenomenon, for example Universitas Global, Fathom.com, or the Global University Alliance.

The global activity of many institutions is visible both in trans-national consortia (Universitas 21, for example, includes 18 research universities from across the world) and in the international reach of individual institutions. The UK's Open University serves 260,000 students in 41 countries and India's Open University (the Indira Ghandi National Open University) licenses material to 11 countries, including the United Arab Emirates, Vietnam and Liberia (Observatory on Borderless Higher Education, Briefing Note, May 2002). Countries are involved in the export and the import of higher education. For example, between 1996 and 2001, Australian off-shore enrolments doubled from 18% to 35% of all international students enrolled in Australian HE institutions. And by 2001, Hong Kong was hosting more than 150 overseas providers of higher education, often in collaborative arrangements with local providers (Olsen, 2002 p.5).

Government policies have played their part both in expanding and regulating public and private sectors of higher education. Governments have also, in many cases, sought to use market forces as a lever for change in public higher education. Policies vary across countries. In some countries, the levers have included expanding the numbers of private and for-profit institutions (as in Malaysia and Kazakstan). In other countries, such as the UK and Australia, policies have included a year-on-year reduction in the unit of resource for teaching in the public universities, with encouragement to institutions to seek alternative funding sources. In other Western European countries, the policy direction has been to increase autonomy in return for accountability and responsiveness (as in the Netherlands and Sweden). In all cases, it is clear that government policies as well as funding and regulatory regimes are important components of institutions' micro-environments. Institutional responses to competition and collaboration are shaped as much by these micro-environmental drivers as by internal institutional drivers such as mission, curriculum and skills' profiles, pedagogical orientations, student demands, market opportunities and partnership options (see Figure 15.1).

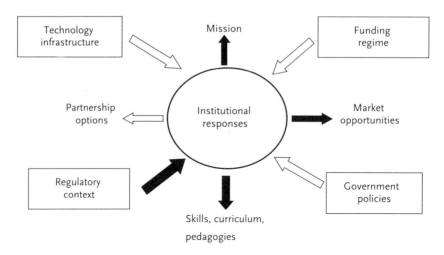

Figure 15.1 Micro-environments and institutional responses

15.3 Regional and National Responses

At regional and national levels, we can observe the interaction between governmental actions and institutional initiatives. In Europe, regional agendas are important both at the Community and sub-regional levels. The European Commission's eLearning Action Plan (2001), part of the wider eLearning initiative (2000) has a number of components that are necessary for the development of institutional strategies. These include the deployment of a high quality technological infrastructure at a reasonable cost, the promotion of digital literacy (particularly among teachers and trainers), creating appropriate conditions for the development of content, services and learning environments, developing relevant standards, and strengthening cooperation and dialogue at all levels including the establishment of public-private partnerships. Member states are using such initiatives to guide and inform their own national policies. In other regions of the world, supra-national organisations such as the World Bank have been involved in helping to establish the conditions for ICT-based learning. The African Virtual University, initially a World Bank project and now an independent non-profit organisation, serves 18 countries in sub-Saharan Africa through multi-mode delivery of courses at undergraduate and post-experience levels. AVU aims to become the architect, facilitator and integrator of an education network that matches student and university needs in Africa with suppliers of content worldwide (www.avu.org). Since its pilot phase in 1997, more than 24,000 students have completed courses in technology, engineering, business and sciences and more than 3,500 professionals have attended executive and management seminars.

In another region, the South Pacific, several governments (Australia, Japan and New Zealand) have collaborated to provide funding for a satellite communications network to serve the 12 member countries of the University of the South Pacific. The USPNET 2000 currently provides high-speed audio services, Internet access, video capabilities and delivery of academic courses to nearly half the student population in the 12 countries (Observatory on Borderless Higher Education, Briefing Note, 2002).

Government-led national eLearning initiatives are visible in several countries. There are both similarities and differences in the rationales and types of initiative. In France, for example, the government is sponsoring 'digital campuses' to boost the provision of on-line courses. Each consortium is discipline-based (specialising in areas such as law, business or engineering); some include foreign partners and all are linked to the National Distance Learning Centre (CNED) although students will enrol with individual universities. The Ministry of Education is hoping to attract traditional students as well as employees wanting to upgrade their professional training and teachers wanting to update their knowledge and skills (Marshall, 2001). The French government's strategy is clearly aimed at maintaining economic competitiveness by responding to the need for lifelong learning including specialised skills, but as Professor Loing tells us, it is also about asserting the French language, culture and educational approach in an area of international competition (Loing, 2002).

Finland's strategy of developing a National Virtual University and National Virtual Polytechnic is also motivated by competitive pressures (a need to ensure that the Finnish university system is protected from outside competitiors) as well as by other rationales. The reported aims of the strategy are to enhance collaborative teaching between universities, promote joint research with industry, preserve national culture and language, promote access to higher education for a widely dispersed population and enable participation in the Information Society through access to training in the use of ICT for all citizens (Karran & Pohjonen, 2001). The collaborative model established is built around regional learning communities involving higher education providers, local and regional agencies and business communities. In Finland's case, the Finnish Committee of University Rectors was a driving force behind the idea of a virtual university and the Ministry of Education provided support to take the agenda forward.

Pakistan's Ministry of Science and Technology is the leading force behind Pakistan's plans to establish a virtual university. In this case, the rationale was somewhat different from the French and Finnish examples, namely a shortage of manpower for the IT sectors in Pakistan, with traditional universities unable to support the demands of the hi-tech sector. The focus of the Virtual University is only on IT education (initially at undergraduate level), with branch campuses across the country. Tutors are to be drawn from both public and private sectors to supplement the full-time faculty. Delivery will initially be through video, moving in the next two years to high-speed computer networks (Del Castillo, 2001a).

In addition to the countries mentioned above, eUniversity initiatives are underway in the UK (e-universities Worldwide), Sweden (the Net University), Malaysia (Universiti Terbuka Malaysia – Unitem), and Greece (the Hellenic Open University) to name but a few. Again, the rationales and forms of these initiatives differ, according to the particular circumstances of the country and its higher education system. The UK's initiative is aimed at increasing the UK's share of overseas students and is principally intended to be a commercial venture. Sweden's initiative, on the other hand, is aimed at making higher education more accessible to remote communities and people at work and at increasing the amount of choice available to students by facilitating the combining of courses from different universities.

The intentions of the Malaysian and Greek governments are different again. In the case of Malaysia's Unitem, a consortium of the country's 11 public universities, established in 1999 as a private company, the aims were several-fold. They included: streamlining the management of distance learning programmes and facilitating the sharing of resources across universities; producing more skilled workers with a science and technology background; offering quality courses more cheaply; increasing the democratisation of education and enabling Malaysia to become a regional education hub (Singh, 1999). In Greece, the Hellenic Open University was established in 2000 to provide access to undergraduate, graduate and doctoral programmes for previously excluded groups of students those without first degrees or those who could not access state universities because of family or work commitments. The University is funded by the Greek government and the European Commission and is proving popular with adult students (Del Castillo, 2001b).

15.4 Institutions' Strategic Choices

In choosing how to position themselves in relation to their environments, institutions face a number of dilemmas. They need to decide whether to engage with ICT developments on their own or in collaboration with other universities, with government or companies, whether they will provide services directly or through brokers and whether they will focus particularly on developing and enhancing programmes or on enhancing services for students (or on both, and in what order of priority). Institutions also need to make choices between for-profit and not-for-profit activities, institution-wide or unit-based initiatives. They need to consider whether greater opportunities lie in providing access to qualifications or to wider learning activities and resources. They will also need to identify their principal aims: to enhance quality and increase choice for existing students or to develop new markets? And will the institution's existing brand and reputation facilitate or impede the exploitation of opportunities or must a new brand be developed?

A recent survey of 500 universities in the Commonwealth undertaken by the Observatory on Borderless Higher Education in May 2002 provides some information about the views and strategies of higher education institutions in relation to ICT-based learning (Observatory on Borderless Higher Education, forthcoming Briefing Note, September 2002). By August 2002, 101 responses had been received (about 20% of the institutions contacted) with 79% from developed countries and 21% from developing countries. The data that follow represent a small snapshot of these responses.

Institutions were asked about their perceptions of the key issues for institutional ICT strategies over the next three years. Their responses suggested that the following are the priority issues:
– Providing on-line provision to campus-based students (50%);
– Providing on-line provision to remote students (41%);
– Providing adequate IT development and support for faculty (61%);
– Integrating academic and administrative IT services and systems (47%).

Outsourcing a greater proportion of IT infrastructure was not perceived as a priority for those that responded to the survey.

Institutions were asked if they had developed institution-wide strategies for on-line learning and if they had, when these had been developed. 40% of respondents reported that they did have such a strategy, 26% were developing one and 16% reported that on-line learning was integrated with other institutional strategies. Most strategies had been developed in the last five years (87%), with a significant proportion (64%) produced since 1999.

Respondents were also invited to comment on the rationales behind their development of on-line learning initiatives. The highest percentage of responses (94% overall) claimed that the main rationales were to enhance on-campus teaching and learning and to improve flexibility for on-campus students (92% overall). Also receiving high response rates were rationales concerning widening access (65% overall and 57% for developing countries) and 'keeping up with the competition' (71% overall and 100% for developing countries). It is unclear, however, whether this is competition from new providers or peer institutions. It is interesting to note that entering new international markets (53% overall and 29% for developing countries) and enhancement of distance learning (59% overall and 71% for developing countries) were also of importance. Other choices received low levels of response overall, for example, cutting teaching costs long-term (20%), although the percentage was much higher amongst respondents from developing countries (57%).

It is interesting to compare these findings with experiences on the ground. A recent report of a study visit to Australian institutions (Boezerooy & Riachi, 2002) made by members of the UK's Association for Learning Technology (ALT) and the Dutch equivalent organisation (SURF-Educatief) notes current trends in Australia:
- An emphasis on using ICT to increase 'flexibility' for students (in terms of ease of administration, delivery options and pedagogical approach) in response to the increasing numbers and diversity of the student population.
- Moves to integrate ICT developments into the mainstream of institutional operations, including integrating support and administrative processes with teaching and learning systems.

Teaching and learning units in many institutions play a leading role in the implementation of the university's ICT strategies in partnership with support from the central administration.
Implementation of e-learning strategies appears often to be made with regard to cost and efficiency savings rather than to particular commitments to improve teaching and learning from a pedagogical basis.

The report concludes that the 13 institutions visited were ahead of many UK and Dutch institutions in their ICT developments. In combination, the study-visit report and the Observatory's survey report may therefore indicate the path that most institutions will be travelling over the next few years. The two reports may also highlight some areas of potential tension, including pedagogical benefits versus cost and efficiency gains.

Conference papers from the recent conference in Rotterdam (2-4 September, 2002) on the 'New Educational Benefits of ICT in Higher Education' (www.oecr.nl/conference) allow some further categorisation of institutional rationales for their ICT strategies (see Figure 15.2). It is important to note that ICT initiatives in institutions are not necessarily *institutionally* driven; student demand for ICT skills (Oblinger & Rush, 1998), teacher innovation (Brown, 2002) and curriculum requirements (Hall, 2002) are each important drivers of innovation and change at departmental, faculty and individual levels. Indeed, at the early stages of introducing ICT into teaching and learning, developments tend to be ad hoc or even serendipitous, as one author has suggested (Devine, 2002). Some commentators (for example, Cooper, 2002) argue that few institutions have yet developed co-ordinated activities for e-learning on an institution-wide basis, while others point to different stages of development in the implementation of e-learning and to success factors that can be identified at each stage (Gans et al, 2002).

263

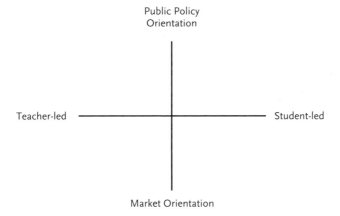

Figure 15.2 Possible institutional rationales for ICT developments

15.5 Corporate and Competitors' Strategic Choices

It is worth pausing for a moment to look outside higher education to consider the kinds of choices that businesses might make in relation to the education and training needs of their business and their consequent relationships with educational providers. Government and institutional responses are only part of the picture in the world of 'borderless education'. Student choices have always been important and increasingly, in a knowledge-driven economy, the kinds of strategies adopted by businesses may also lead higher education in particular directions.

The strategic choices made by businesses are likely to be influenced by their perceptions of the value of education and training to their own core business. They may also assess the relative benefits of providing education and training in-house or by outsourcing provision to education providers (whether public or for-profit providers). Figure 15.3 illustrates the kind of choices businesses may make.

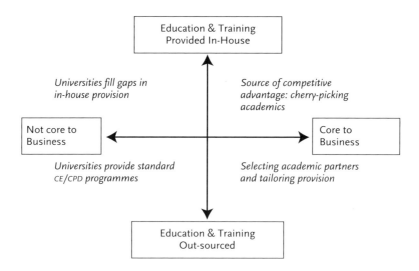

Figure 15.3 Choices companies may make

If education and training is viewed as core to business, the choice may be to provide it in-house, in which case academics and other professionals could be employed directly to design relevant internal programmes, perhaps through the vehicle of a corporate university. An alternative solution would be to outsource provision to universities and colleges (who would then remain the employers of the educational staff), ensuring that provision was tailored to the needs of the business. If education and training is not seen as core to business, the relationship to higher education is different. Institutions could either provide generic modules to supplement in-house provision or offer traditional undergraduate and postgraduate programmes and continuing professional development. ICT makes it possible for higher education provision to be much more closely integrated with business needs, both through tailored content and through flexibility in modes of delivery.

Another point of reference are the strategies and choices made by potential competitors (such as the successful for-profit universities and educational businesses). Engagement by institutions with these companies as partners and observation of their actions as competitors illustrates the kinds of strategic choices that they are making. First, the commercial viability of programmes and curricula is of critical importance, both in terms of return on investment for the purchasers and in terms of profitability and market share for the suppliers. For the for-profit suppliers, commercial viability is achieved through provision that is relevant, of high quality, accessible and cost-effective (Cunningham *et al.*, 2000).

265

A second choice concerns delivery locations. It is noticeable that some of the large US players (such as the Apollo Group, Sylvan Learning Systems and ITT Educational Services) are becoming increasingly international in their reach. Apollo International has operations in Puerto Rico, Canada and the Netherlands, with new developments emerging in Brazil and India (Olsen, 2002). Sylvan has expanded through part-acquisitions of overseas' private universities, or full ownership where national regulations allow. Sylvan Universities International now has on-campus programmes in Chile, Spain, Switzerland and most recently, in France. Like Apollo, it too has made an approach in India (Ryan, 2002). ITT Educational Services, a leading for-profit degree-granting institution in the US that specialises in IT and electronics degrees, has also recently entered the Indian market. The company is partnering with NIIT, one of the world's largest software services and ICT education business, with headquarters in India. The partnership with ITT Education Services gives NIIT access to international degree-granting powers in return for development of the technical infrastructure to deliver on-line degrees (Observatory on Borderless Higher Education, Breaking News Bulletin, 12.8.02).

A third issue concerns collaboration; what advantages can be gained by collaborating with institutions and other companies? The benefits include the leverage that can be obtained for the education business' own products and services, access to accreditation for the provision of qualifications and access to local knowledge in overseas' locations including understanding of markets and the regulatory context. Finally, the for-profit education businesses are making choices about forms of collaboration. These include joint ventures, strategic alliances and increasingly, investments and acquisitions. For example, Pearsons and Thompson Learning have entered into strategic alliances with institutions, joint ventures with Learning Technology Platform vendors, have acquired assessment and testing companies and have invested in e-Libraries. Through these varied relationships, the organisational forms of 'learning services' are converging and changing.

15.6 Implementation Issues for Institutions

There is a growing literature on the kinds of issues that institutions face in implementing ICT strategies for teaching and learning, research and administration, with many case study examples available. Papers highlight the strategy development process, including both conceptual and practical aspects, issues of structure and approach, technical decisions concerning the choice of technology infrastructure and administrative systems, as well as fundamental issues of quality and cost. A number of conceptual frameworks have been developed to illustrate the management issues that institutions must attend to in implementing their strategies. Two illustrations are offered below (Figures 15.4 and 15.5). The first framework originated in MIT, but was adapted by Australian researchers investigating how institutions were managing the introduction of technology to deliver and administer higher education (Yetton, 1997). The second is also taken from Australian research (McNaught & Kennedy, 2000), this time focusing on the factors that supported the adoption of computer-facilitated learning within Australian universities. The second framework emphasises the three major themes of policy, culture and support, highlighting areas of overlap between and within themes. Together, these frameworks illustrate both 'hard' and 'soft' issues of implementation.

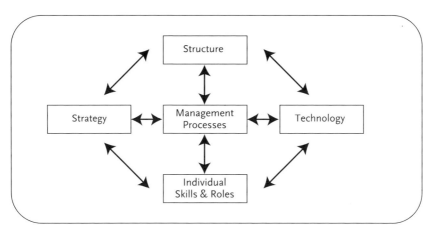

External Environment

Figure 15.4 Organisational framework for implementing ICT strategies (source: Yetton, 1997)

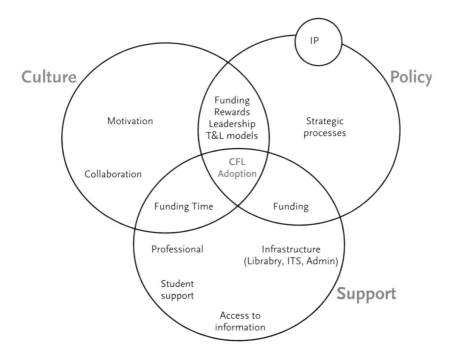

Figure 15.5 Organisational themes affecting the introdution of computer-facilitated learning (Source: McNaught & Kennedy, 2000)

Time and again, studies point to issues of:

- Coherence and integration of policy across institutional operations;
- Leadership, with visible and committed support at all levels;
- Staff attitudes, rewards and recognition, training, and the development of new roles and relationships;
- Quality assurance and IPR management;
- Student learning and lifestyle issues;
- Resourcing: technological, human and financial;
- Processes of managing change.

Studies also provide valuable information about the particular lessons learned through developing and implementing institutional strategies, whether these are local and regional or international. One recent case study incorporating such lessons of experience (Cooper, 2002) provides considerable detail from one institution that is seeking to address all three geographical markets. The author's conclusion is that 'e-learning and borderless education must not be ghettoised', but must become part of the cultural mainstream of the institution. To this end, 'a strong, corporate, partnership approach (including in this case a wide range of local and international partners) can be pivotal to success'. (Cooper, 2002: p.25)

15.7 Benefits and Impact

To complete this tour of institutional strategic responses to ICT developments, it is useful to consider the outcomes and impact of such responses.

Addressing pedagogical issues, authors have pointed to benefits in terms of greater collaboration among students, opportunities to involve practitioners as well as scholars with students (Brown, 2002) and opportunities for students and staff to develop new skills and competencies (Wielenga, 2002). Benefits have also been assessed in terms of the quality of student learning and the impact of ICT on student lifestyles. These types of benefit and impact include levels of student motivation and responsibility for their learning (Timmis & Cook, 2002) and the reduction of cultural boundaries in management education (Blake & Go, 2002).

The impact of ICT on academic and administrative staff, particularly in terms of attitudes towards innovation and approaches to teacher training are also topical. Several institutions have developed central units that offer a range of resources and support for staff and educational development and the work of these units is discussed in case studies.

Papers also focus on the impact of ICT on in-company training, particularly in terms of flexibility for employers and employees (Gelderblom, 2002), while others point to the benefits accruing from partnerships between higher education institutions and companies in the evolution of e-Learning networks (Kuchler & Kayser, 2002). Reported benefits include cost-reduction, risk-sharing and transfer of technical expertise.

Finally, where papers focus on strategy implementation, they address questions of organisational impact, particularly in terms of structure and culture. In looking towards the future for higher education institutions, these issues are of particular interest since they begin to throw light on different images of universities and colleges as organisations. Three examples are offered as emergent organisational images.

The first set of images comes from Yetton's Australian work (Yetton, 1997). He identifies three structural forms that, although not yet clearly specified, illustrate different ways of managing the introduction of ICT across institutional activities and of competing in a 'borderless world'.

269

a The 'Old' University or Professional Bureaucracy

Here ICT is used to enrich an elite learning community by funding independent new ventures while maintaining the established base. Successful ventures based on different and unique competencies grow and 'feed' the established campus with innovations in teaching, learning and research. The ventures have more freedom to select, contract and reward expertise, thus attracting academic entrepreneurs and risk-takers and encouraging flexibility of administrative systems.

b The 'Divisional University'

ICT in this case supports the success of semi-autonomous faculties. The devolved faculties are enabled by a powerful central ICT infrastructure and each faculty has different competencies and strategic foci. This form is more complex to manage given its scale, but relies on the academic divisions to manage some of the complexity. Faculties have their own ICT support, management processes, specialist roles and skills which focus on their own particular areas of applied research and teaching.

c The 'New' University

ICT in this case is central to and critically underpins the strategic agenda. A new 'subsidiary' delivers ICT based teaching and learning, developing innovations and new core competencies in a separate, centrally resourced unit. In such a 'greenfield' site, highly skilled experts can be selected as required, with a focus on the motivation and ability to work in multi-functional teams.

Yetton argues that institutions are more suited to one of these models than to the others for reasons of history, size, age and reputation. They therefore need to choose a strategic focus that fits their macro and micro-environments.

A second image is taken from a research project in the UK entitled 'Space, Place and the Virtual University' (Cornford, 2000). The findings of this research project suggest that, perhaps paradoxically, the concept of the 'virtual university' as the epitomy of flexibility and 'borderlessness' is in reality somewhat removed from that image. The concept of the virtual university, 'a university without walls', implies the removal of barriers of time (through asynchronous learning), of space (through study any-place) and of geography (through transnational opportunities). In reality, as Agre (2000) points out, 'information and communications technologies create incentives to standardise the world'. The UK project, which focused in depth on a small sample of four different institutions, found numerous examples of how the introduction of ICT into teaching, learning and associated administrative systems led to increased needs for centralisation, stream-lining and co-ordination. The authors comment that

> the process of trying to apply ICTs in the context of more or less traditional structures appears to generate demands for the construction of a new institu-tional structure capable of co-ordinated action with formalised roles and standardised practices into which technologies can be fitted and in which they can operate. These demands literally call into being a different, and more concrete, type of institution. (Cornford, 2000: p.19)

Cornford suggests in terms of organisational image that ICT is pushing on-campus universities towards a more corporate form, characterised by tighter policy definition and implementation. (On the other hand, distance-education institutions would already recognise the requirement for formalised systems and structures, see Daniel, 1998). But, the compelling need for 'enterprise' in institutions arising from the macro-environmental pressures described earlier, may require institutions to move towards different organisational forms, perhaps like Morgan's 'spider plant' image (Morgan, 1997: p.63); indeed, the latter seems to resemble some of the new forms of 'learning businesses'. Cornford acknowledges a potential tension between 'the corporate' and 'the enterprise' university.

Where the two previous examples were drawn from research on existing institutions, the third example was developed as a prototype during a research fellowship with the British Telecom company (Squires, 2000). The basic concept underlying this image is that ICT can free higher education teachers from their institutions since they can act as free-lance agents (and indeed, free institutions from some of their staffing constraints). The concept that Squires develops of a 'Peripatetic Electronic Teacher' (PET) draws on traditions both ancient and modern. From ancient times we have the example of 'the wandering scholar'; in the US of today the nine-month teaching contract already necessitates some free-lance work and, most recently, there are the consultancy possibilities opened up by companies such as Hungry Minds.com. The latter company offered the opportunity for individual academics to advertise their products and services through a network of personalised websites. Squires' prototype draws on the continuing work of Brown and Duguid, begun in 1995, on 'Universities in the Digital Age'.

271

Squires describes the creation of a 'PET- World', a networked environment that acts as a gathering place or 'agora' that brings people together, encouraging participation and creativity. This world is inhabited by PETs and learners. Squires developed a design specification for a PET-World based on four 'presence domains':
- *Pedagogical presence* where the PET appears as a teacher, playing roles such as instructor, coach, mentor, tutor and expert;
- *Professional presence* where the PET appears as a member of the teaching profession, playing roles such as colleague, committee member, trainer and trainee;
- *Commercial presence* where the PET appears as a free-lance worker available for hire, playing such roles as consultant, personal tutor and publisher;
- *Managerial presence* where the PET appears as an administrator, scheduling teaching commitments, validating learners' attendance and achievements and managing course enrolments.

Each of these domains has further dimensions, for example, the Pedagogical Presence Domain includes a Learning Theatre, Discourse Forum, Virtual Field Centre, Learning Surgery, Design Studio and Resource Centre.

Some of these apparently speculative ideas have in practice already taken root in the design of virtual and corporate universities, knowledge guilds and networked communities of practice. In building real models out of such prototypes, organisational images of higher education are expanded with potentially interesting results and unforeseen benefits. When focusing on the practical and policy issues that institutions face as they seek to devise and implement their ICT strategies, we should not lose sight of the potential impact of developments on our images of universities and colleges as they are, or might be. Institutions and individuals alike need to face the challenges and opportunities of ICT armed both with the lessons of experience and evidence that research can provide and with the creativity that comes from imaginative speculation on future possibilities.

References

Agre, P. 2000. Infrastructure and Institutional Change in the Networked University. *Information, Communication & Society.* Vol. 3, 4, pp.494-507.

Blake, C. & Go, F. (2002). Stimulating e-Learning in Europe: A Supply Chain Approach. *Paper presented at the European Conference: The New Educational Benefits of ICT in Higher Education.* Rotterdam, 2-4 September, 2002.

Blight, D. (1995). *International Education: Australia's Potential Demand and Supply.* Canberra, IDP Education Australia.

Bohm, A. & King, R. (1999). *Positioning Australian Universities for the Future: An Analysis of the Education Markets in the People's Republic of China.* Canberra, IDP Education, Australia.

Boezerooy, P., & Riachi, R. (2002). Keeping up with the Neighbours: e-Learning in Australian Higher Education. *Key Findings from the surf/ALT Tour. Paper presented at a Generic LTSN/ALT Workshop at Manchester Metropolitan University,* 9 July, 2002.

Brown, D. 2002. Proven Strategies for Teaching and Learning. *Paper presented at the European Conference: The New Educational Benefits of ICT. Rotterdam, September 2-4, 2002.*

Brown, J. S. & Duguid, P. (1995). 'Universities in the Digital Age', Work in Progress University Paper, http://www2.parc.com/ops/members/brown/papers/university.html (Retrieved 23.9.02).

Collis, B. & Gommers, E.M. (2001). Stretching the Mold or a New Economy? Part 1: Scenarios for the university in 2005. *Educational Technology,* Vol. XLI (3), pp. 5-18

Cooper, A. 2002. Barriers, Borders and Brands: Forging an Institutional Strategy for Development and Collaboration in Borderless Higher Education. *Paper prepared for the Observatory on Borderless Higher Education. May 2002*, www.obhe.ac.uk

Cornford, J. 2000. The Virtual University is....the University made Concrete? *Information, Communication & Society*, Vol. 3, (4), pp. 508-525.

Cunningham, S., Tapsall, S., Ryan, Y., Stedman, L., Tapsall, S., Bagdon, K.& Flew, T. (1998). *New media and borderless education: A Review of the convergence between global media networks and higher education provision*. Canberra, ACT: Department of Education, Training and Youth Affairs.

Cunningham, S., Ryan, Y., Stedman, L., Tapsall, S., Bagdon, K., Flew, T. & Coaldrake, P. (2000). *The Business of Borderless Education*. Canberra, ACT: Department of Education, Training and Youth Affairs.

Committee of Vice Chancellors and Principals (CVCP, now Universitiesuk). (2000). *The Business of Borderless Education: UK Perspectives*. (Vos1-3). London: author.

Daniel, J. (1998). *Mega-Universities and Knowledge Media*. London, Kogan Page.

Del Castillo. (2001a). September 20. Pakistan Plans its First Virtual University. *The Chronicle in Higher Education*. http://chronicle.com. Retrieved 20.9.01

Del Castillo. (2001b). August 10. Longtime Professor Helps Guide Greece's New Distance Learning Institution. *The Chronicle of Higher Education*. Http://chronicle.com. Retrieved 10.8.01

Devine, J. (2002). ICT for Teaching and Learning – Strategy or Serendipity? – the Changing Landscape in Ireland. *Papre presented at the European Conference: The New Educational Benefits of ICT in Higher Education*. Rotterdam, 2-4 September 2002.

European Commisson (2001) *The eLearning Action Plan: Designing Tomorrow's Education*. Brussels, 28.3.2001. COM(2001) 172 final. http://europa.eu.int/information_society/eeurope/action_plan/eeducation/index_en.htm

Gans, R., van Hoff & P., Basoski, I. (2002). The implementation of e-Learning in Higher Education: Facts, Figures and Future. *Paper presented at the European Conference: The New Educational Benefits of ICT in Higher Education*. Rotterdam, 2-4 September, 2002.

Gelderblom, A. (2002) ICT: New Opportunities for Higher Education Institutes to Train Employees? *Paper presented at the European Conference: The New Educational Benefits of ICT in Higher Education*. Rotterdam, 2-4 September 2002.

Hall, R. (2002) Aligning Learning, Teaching and assessment using the Web: an evaluation of pedagogic approaches. *British Journal of Educational Technology*. 33, 2, 49-158.

Karran, T. & Pohjonen, J. (2001). National Virtual University Finland: First steps to creating a virtual university, some preliminary questions and possible answers. *Proceedings of European Association of Distance Teaching Universities (EADTU) Millenium Conference: Wiring the Ivory Tower, Linking Universities Across Europe* (: 161-164). Heerlen, the Netherlands: EADTU Secretariat.

Kuchler, T. & Kayser, S. (2002). CLIX Campus and the imc Higher Education e-learning Network: A Private Public Partnership-Approach to Creating New Educational Benefits. *Paper presented at the European Conference: The New Educational Benefits of ICT in Higher Education.* Rotterdam, 2-4 September, 2002.

Loing, B. (2002). The French approach to e-Learning: Public initiatives for virtual campuses and a Francophone medical school. *Paper presented at the European Conference: The New Educational Benefits of ICT in Higher Education.* Rotterdam, 2-4 September, 2002.

Marshall, J. (2001), August 3. Six online pioneers launched by French. *The Times Higher.* 12.

McNaught,C. & Kennedy, P. (2000). Staff Development at RMIT: bottom-up work serviced by top-down investment and policy. In Squires,D., Conole,G., & Jacobs,G., (eds) (2000). *The Changing Face of Learning Technology.* Association for Learning Technologies, Cardiff, University of Wales Press.

Morgan, G. (1997). *Images of Organization.* London, Sage

Newman, F. & Couturier, L. (2002). Trading Public Good in the Higher Education Market. *Paper prepared for the Observatory on Borderless Higher Education.* January, 2002, www.obhe.ac.uk

Oblinger, D. & Rush, S. (1998). *The Future Compatible Campus.* Boston, Anker.

Observatory on Borderless Higher Education, Briefing Note, 4, May, 2002, www.obhe.ac.uk

Observatory on Higher Education, 'Breaking News', 12.8.02, wwww.obhe.ac.uk

Observatory on Borderless Higher Education, 'Breaking News', forthcoming, September, 2002, www.obhe.ac.uk

Olsen, A. (2002). E-Learning in Asia. *Paper prepared for the Observatory on Borderless Higher Education.* June, 2002, www.obhe.ac.uk

Ryan, Y. (2002). Emerging Indicators of Success and Failure in Borderless Higher Education. *Paper prepared for the Observatory on Borderless Higher Education.* February, 2002, www.obhe.ac.uk

Singh, S. (1999), August 6. Open University to begin with intake of 257. *New Straits Times*, p. 5.

Squires, D. 2000. Peripatetic Electronic Teachers in Higher Education. In Squires,D., Conole, G., & Jacobs, G., (eds). (2000). *The Changing Face of Learning Technology.* Association for Learning Technologies, Cardiff, University of Wales Press.

Tate, W. (2000) *Implications of Futures Studies for business, organisation, management and leadership.* [Report]. London: Council for Excellence in Management and Leadership.

Taylor, S., & Paton, R. (2002). Corporate Universities: Historical development, conceptual analysis and relations with public sector higher education. *Paper prepared for the Observatory on Borderless Higher Education.* July, 2002, www.obhe.ac.uk

Timmis, S. & Cook, J. (2002). Motivating Students towards On-line Learning: Institutional Strategies and Imperatives. *Paper presented at the European Conference: The New Educational Benefits of ICT in Higher Education.* Rotterdam, 2-4 September, 2002.

Uscensusbureau. (2002). IDB Summary Demographic Data. www.census.gov/cgi-bin/ipc/idbsum?cty=RP. Retrieved 30.8.02.

Van Damme, D. (2002). Quality Assurance in an International Environment: National and International Interests and Tensions. *Background Paper for the CHEA International Seminar III, January 24th, 2002, San Francisco.*

Wielenga, D. (2002). Probing and Proving competence. *Paper presented at the European Conference: The New Educational Benefits of ICT.* Rotterdam, 2-4 September, 2002.

World Bank. The Task Force on Education and Society. (2000). *Higher Education in Developing Countries: Peril and Promise.* Washington, DC: 107, 111. www.avu.org

Yetton, P. (1997). *Managing the Introduction of Technology in the Delivery and Administration of Higher Education.* [Report] Canberra, act: Department of Education, Training and Youth Affairs.

Author's note

The author is grateful to colleagues at the Observatory on Borderless Higher Education (jointly established by the Association of Commonwealth Universities and Universities UK) for their assistance with some of the data in this paper. Richard Garrett, Research Officer at the Observatory, deserves particular thanks.

16 Proximity and Affinity: Regional and Cultural Linkages between Higher Education and ICT in Silicon Valley and Elsewhere

Hans Weiler, Stanford University, United States of America

16.1 A Personal Preface

The subject of this chapter, if I may start on a personal note, carries a particularly strong personal connotation inasmuch as it reflects on an important part of my own biography. When I first came to Stanford in 1965 as a young assistant professor, there was no 'Silicon Valley' yet, that name wasn't invented until the early 1970s, but my letter of appointment carried the signature of Frederick Terman, then the provost of Stanford University. It was Terman who, just a few years earlier, had started, together with his former students William Hewlett and David Packard, the 'Stanford Industrial Park', which was the nucleus of what was to become Silicon Valley. Over the next thirty years, that development occurred virtually on our doorstep: from my house at Stanford the main entrance of Hewlett-Packard is literally around the corner.

In the late 1970s, one of the great treats for our children was to ride the famous roller coaster at the 'Great America' amusement park near San Jose, around which the traditional orchards were rapidly being replaced by high-tech companies. As it turns out, the roller coaster serves as a pretty good metaphor for the bust-and-boom cycles that Silicon Valley has gone through over the past 30 years.

And when, in 1993, I went from Stanford to Germany to assume the presidency of Viadrina European University in Frankfurt/Oder, Intel had just introduced the fifth generation of its microprocessors, called Pentium, just next door, incidentally, from those roller coasters that our children used to ride. [One of those roller coaster-riding children, by the way, now a professor of labour economics and regional studies at Colorado State, deserves special credit for my continuing education in regional studies and, hence, for this paper (Weiler, S. *et al.*, 2001).]

But the history of Silicon Valley has been a roller coaster in other respects as well. We bought our house at Stanford in 1972 for $56,000; that house would now cost well over a million dollars to buy. That would be good for us, if ever we wanted to move to Nebraska, but terrible for people who now seek affordable housing in Silicon Valley, including young assistant professors. And when we returned to Stanford from Germany in 1999, we were appalled to see what the newest boom of Silicon Valley in the 1990s had done to the traffic gridlock on the streets of Northern California.

I mention all of this at the outset to dispel any notion that I am presenting myself here as the unreconstructed enthusiast of the Silicon Valley experience. That experience is decidedly mixed. But I do regard what happened in this Valley over the last thirty years as a remarkable phenomenon that is very much in need of better understanding, because of both its successes and its failures, and especially because of the sometimes rather naïve attention it has attracted internationally.

278

16.2 Overview

Nowhere in the world is the interaction between higher education and its regional context in the realm of technology as highly developed, as thoroughly analysed and as widely commented upon (and often criticised) as in Silicon Valley. The question of how this, even by American standards, remarkable degree of symbiosis has been possible has found many answers in what is by now a fairly respectable literature on the subject.

Some of these answers focus on the very particular corporate culture that has emerged in Northern California's high-tech industry and on its peculiar mixture of competition and cooperation. Other answers have emphasised some of the characteristics of the North American academic culture in general, and of the special configuration of higher education, research, and development in the San Francisco Bay Area, in particular.

A full understanding of the unique technological and entrepreneurial environment that has emerged in this region over the last 40 years requires, however, a more encompassing view of how these two cultures, the corporate culture of the Silicon Valley high-tech industry and the academic culture of institutions like Stanford and Berkeley, complement and draw on each other. This mutual relationship is highly localised; its success depends on the physical proximity of its partners within a narrowly defined geographical space.

Thus, in the kind of symbiotic relationship that has developed between higher education and high-tech industry in Silicon Valley, proximity clearly matters. But that is not the only thing that matters: proximity is a necessary, but not a sufficient, condition for successful symbiosis. What also matters is a basic affinity between the corporate and the academic cultures that are part of this relationship, the sharing, in other words, of certain traits between knowledge-based enterprises and entrepreneurially oriented institutions of higher learning and research. There clearly is a 'regional advantage', as Annalee Saxenian (1996) shows in explaining the superior performance of Silicon Valley in comparison with other American high-tech regional clusters (such as the one that developed around Route 128 in Massachusetts). But this regional factor has to be complemented by the kind of 'cultural advantage' that the basic affinities between higher education and corporate technology provide.

This chapter explores both of these dimensions, proximity and affinity, in an attempt to pull together what we know by now about the reasons for the successes as well as some of the failures of Silicon Valley. Against this background, the chapter also raises the question of whether, and under what conditions, similar symbiotic relationships could emerge in other parts of the world, notably in Europe. The thesis that I advance here answers this question with considerable scepticism. This scepticism is based in part on the difficulty of reproducing the 'regional advantage' of Silicon Valley in Europe, but more importantly on the considerable lack of affinity between European institutions of higher education and research, on the one hand, and technological entrepreneurship, on the other. Notable recent developments in this regard notwithstanding, this chapter maintains that the gap is still rather wide.

The focus in this analysis is on information and communication technology (ICT) which is by far the largest part of the production spectrum in Silicon Valley, especially when one includes the ICT components and infrastructures that other high-tech developments, notably in biotechnology or medical technology, require.

279

The analysis starts out with a review of the argument that proximity matters, and that the close configuration of high-tech companies, universities and other research institutions has made possible a density of interaction rarely found elsewhere.

I then proceed to a study of some of the cultural traits of Silicon Valley's corporate sector, where the interaction of certain structural conditions, a special brand of entrepreneurial personality, and the institutionalisation of particular norms and patterns of behaviour have produced an environment *sui generis* which has proved unusually supportive of innovation and change.

In a third section, the chapter focuses on some particular characteristics of U.S. higher education institutions, and especially of universities like Stanford and Berkeley, that resonate particularly well to, and interact especially well with, the corporate culture of the region.

I conclude, fourthly, by showing why, against the background of this analysis, it appears to be so difficult to replicate both the regional and the cultural advantage of Silicon Valley elsewhere, and in what direction one might seek and find possibilities of replication.

One other note: When one speaks of the 'success' of Silicon Valley, the real story is one of considerable ups and downs, of cycles of boom and bust. The history of Silicon Valley over the past thirty years provides an instructive series of such cycles, each of which has resulted in a substantial loss of jobs, but which has also each time generated a new wave of invention and innovation. This was true of
- the recession resulting in the early 1970s from cutbacks in defence spending, which led to exploration of the commercial applications of defence technologies;
- the recession in 1985 resulting from over-capacity in the semiconductor industry, which led to a concentration on higher-value microprocessors;
- the recession in 1990 resulting from over-capacity in the personal computer industry, which led to the development of the Internet.

The most recent recession, which began in 2000 with the bursting of the Internet and dot.com bubble and was exacerbated by the fallout of the events of September 11, 2001, is once again weighing heavily on the fortunes of Silicon Valley. Analysts predict that, here again, the current bust will marshal innovative energies for a new boom that is likely to be directed to such fields as the mobile Internet and wireless communication, new applications of technology in education and elsewhere, further advances in biotechnology (bioinformatics, biomaterials, biochips) and the field of nanotechnology (The Next Silicon Valley, 2001, 8-11; Rowen 2000, 198-199).

16.3 Proximity Matters: the Regional Advantage of Silicon Valley

Ed McCracken, Chairman and CEO of Silicon Graphics, Inc. once explained the importance that his company attaches to regional proximity:

> We drew a ten-minute commute circle around Hoover Tower [on the Stanford campus] to define acceptable locations for our company. (Source: Gibbons 2000, 213).

Commuting has become much more difficult in Silicon Valley in recent years, but the principle still holds. Indeed, the very nucleus of Silicon Valley, Stanford Industrial Park, was conceived, and succeeded, on the basis of the physical proximity between the first high-tech start-ups of the 1950s and 1960s, notably Hewlett-Packard, Varian, and Fairchild, and the laboratories, libraries and lecture halls of Stanford University. They were not even ten minutes of commuting, but merely a short bike ride away from one another. The possibility for the new companies to use Stanford laboratory facilities for their development work, and the opportunities for Stanford students and graduates to work in high-tech companies at the cutting edge of technological development proved to be such a fruitful kind of symbiosis that it has found numerous replications up and down the valley that stretches between the Stanford campus and San Jose Airport.

The elements of this symbiosis have multiplied since those early initiatives; all of the many and expanding opportunities of electronic and remote communication notwithstanding; however, virtually all of them have benefited from the physical closeness between institutions of higher learning, research institutions, and high-tech companies. Looking a bit more closely at this symbiosis leads me to the following observations:

a   A particularly important role was played, and continues to be played, by university-based programmes of continuing education for the engineers and scientists of cooperating companies, beginning with the 'Honors Cooperative Program' initiated by Stanford provost Frederick Terman in

1953, which combined classroom instruction on the Stanford campus with instruction via close-circuit instructional television links to individual companies (Saxenian, 1996: 23), and continued by its modern day successor, the Stanford Center for Professional Development that now has 452 companies as members and uses a mix of virtually all available technologies in delivering credit courses, industry seminars, and professional continuing education programs (DiPaolo 2002).

b Stanford's most valuable and generous resource was, and still is, its land. Considerable tracts of university land were allotted for the creation of the Stanford Industrial Park in the 1950s, where companies such as Varian, Hewlett-Packard, General Electric and others benefited both from attractive leasing conditions and the proximity of Stanford's intellectual resources (Castilla *et al.*, 2000: p. 230).

c As Silicon Valley got off the ground, the opportunities in the region for Stanford students, graduates and professors to form their own companies while retaining their affiliation with the university multiplied. Examples include Hewlett-Packard, Sun Microsystems, Yahoo and many others; defining 'Stanford start-ups' rather tightly as companies where 'both the technology for the first product and a majority of the founding team came from Stanford', Gibbons (2000: p. 202) has calculated that about 60 percent of Silicon Valley revenue in both 1988 and 1996 was produced by Stanford start-ups (Gibbons, 2000: pp. 204-205).

d Another element in the linkage is the growing importance of the licensing and patenting of faculty inventions both for the financing of universities and for fostering the links between university and industry (Grindley & Teece, 1997; Henderson *et al.*, 1998); the Chronicle of Higher Education (2002) has estimated, using data from the last five years, that Stanford, leading all research universities in the U.S., generates licensing income of eight cents for every dollar of research spending, followed by the University of California (6 cents), the University of Wisconsin at Madison and the University of Washington (4 cents each), and MIT and the State University of New York (3 cents each).

e One of the newer developments in this symbiotic relationship is the emergence of new hybrid types of institutional cooperation between universities and high-tech industry, exemplified by the Center for Integrated Systems, 'a partnership between Stanford University and member industrial firms to produce world class research and Ph.D. graduates in fields related to integrated systems', where the 'member companies provide financial support, interaction with their engineers and scientists, and access to resources (such as software, hardware and fabrication, etc.). Through participation by their senior executives, they also advise and guide research directions and curriculum development. CIS programmes are managed through the guidance of Stanford faculty, using their standards for academic achievement.' (from the Center's website, www-cis.stanford.edu); membership is $150,000 a year, and the list of members reads like a Who's Who of Silicon Valley. Institutions like the CIS serve an increasingly important brokering function at the interface between corporate and academic interests (Castilla *et al.* 2000: pp. 229-233), what Hirsch (1972) has called 'boundary-spanning units' designed to connect different institutional worlds; not surprisingly, their role has also become the subject of considerable criticism regarding the danger of undue influence on the university's freedom of research (Noble 2001; Aronowitz 2000, pp. 43-44; Press & Washburn 2000).

f   Silicon Valley has seen a particularly high degree of mobility of scientists between academic and corporate roles, holding these roles often at the same time and with the tacit or open consent of the university on the grounds that these linkages provide excellent avenues for both effective technology transfer into the industry and for alerting the university to new demands and opportunities in technological development.

So much for some of the more important linkages in this network. Even though some of them might work, and have worked, over larger distances, they have proven to be particular effective in the Silicon Valley context by virtue of the physical proximity of the participants. What Castells (1996) has called a 'new spatial logic' has demonstrated its particular strength in the proximity of the networks that have emerged in Silicon Valley, and has provided new answers to the question of 'why clusters cluster' (Brown & Duguid, 2000: p. 17). In reviewing what they call the 'mysteries of the region', John Seely Brown and Paul Duguid of the Palo Alto Research Center of Xerox emphasise 'the character of the local and the importance of direct human interaction', especially where, as in information and communications technology, knowledge is a critical factor (Brown & Duguid, 2000: p. 19): '... learning, innovating, sharing practices, and circulating inchoate knowledge all require reciprocity, close interaction and mutual exchanges among the people involved', and they see 'the workings of the universities within the region' as a prime example of this kind of reciprocity (Brown & Duguid, 2000: p. 35).

Just recently, *The Economist*, in its August 24, 2002, issue, joined this argument by claiming that 'physical presence counts even more than it used to', all the possibilities of modern remote communication notwithstanding,

283

and cites Silicon Valley as one of its prime examples as it concludes: 'One of the mysteries of the wired (and wireless) world is that proximity still counts.' (*The Economist*, 2002: p. 50)

16.4 The Entrepreneurial Culture of Silicon Valley

There is something about the corporate culture of Silicon Valley that sets it apart not only from the corporate world outside of the United States, but from other industrial regions in the U.S. as well. Annalee Saxenian of Berkeley makes a rather compelling case for this essential difference in values and behaviour in her comparison between Silicon Valley and Route 128, the major technological development circling the greater Boston area in Massachusetts. She points out the 'complex balance of cooperation and competition' (Saxenian, 1996: p. 149) which has made possible a much greater openness across corporate boundaries and has meshed well with a prevailing corporate model of relatively open, decentralised, and specialized structures; 'some secrets are more valuable when shared' (Lee *et al,.* 2000: p. 10) is a typical attitude in Silicon Valley. It is this interaction between cultural and structural openness which Saxenian and others see as the principal reason for the distinct advantage of Silicon Valley over Route 128 in recovering from the crisis of the 1980s. This can be documented particularly well in a comparison of corporate cultures and structures between Sun Microsystems and Hewlett-Packard on the Silicon Valley side and Apollo and Digital Equipment Corporation (DEC) in Massachusetts (Saxenian, 1996: p. 126ff.).

Beyond this broad contrast of corporate cultures, there are a number of more specific characteristics of the Silicon Valley setting that play in my judgment an important role in explaining its unique development. The following strike me as particularly salient (see also Weiler, 1998):

a   Entrepreneurial failure is not seen as a sign of defeat, but as a valuable learning experience (Gibbons, 2000: p. 211): 'Silicon Valley is quick to forget mistakes' (*The Economist*, 1997: p. 8) and 'it's hard to learn when you succeed' (*Business Week*, 1997: p. 146). This 'tolerance of productive failure' (Gibbons, 2000) is perhaps one of the key cultural traits of Silicon Valley, and the one hardest to replicate in systems where bankruptcy is still seen as a major and fatal individual and corporate catastrophe.

b   Risks are sought and accepted to a rather unusual degree, which among other things explains and makes possible the particular culture of venture capitalists. This culture, as one analyst has it, proceeds on the calculus that, out of 20 companies, four will go bankrupt, six will stay in business but lose money, six will produce a modest return on the investment, three will do pretty well and one will 'scoop the jackpot' (*The Economist*, 1997: p. 11).

c   There is a major cultural commitment to change, as reflected in the maxim: 'Either we obsolete ourselves, or the competition will' (*The Economist*, 1997: p. 11). The result, both for individual companies and for the region as a whole, is a spirit of constant experimentation and a sense that 'Silicon Valley continues to reinvent itself' (Saxenian, 1996: p. 161), much more so than its competitors in other regions of the U.S., and one of the reasons why Silicon Valley has always managed to bounce back from its various defeats.

d   While the reinvestment of profits to assure further growth is standard business practice everywhere, it has become a particularly focussed practice in Silicon Valley. This is not only reflected in the substantial contribution of high-tech companies to the venture capital pool of the region, but also in the considerable investments by Silicon Valley companies in the training and research capacity of their partner institutions in higher education. Stanford University has been a particularly fortunate beneficiary of this strategy: the hundreds of millions in gifts made to Stanford by the Hewlett and Packard families alone, in inflation-adjusted dollars, are said to rival the founding bequest made by Leland and Jane Stanford at the university's inception in the 1890s (Kaplan, 1999: p. 37). But all higher education institutions of the region, including the junior colleges, have benefited substantially from this corporate strategy of reinvesting into the Valley's infrastructure.

285

e   One honors achievement, and nothing else. As Steve Jobs, one of the founding fathers of Apple, once put it succinctly: 'What matters is how smart you are.' This has had, among other things, the interesting result of opening up the Silicon Valley labour market to immigrants in major ways. One third of the engineering workforce in Silicon Valley comes from mainland China and India, and immigrants play an increasingly important role at the entrepreneurial level as well, beyond such well-known foreign-born members of the founding generations as Andy Grove of Intel who came from Hungary, Eric Benhamou from Algeria at 3Com, Philip Kahn from France at Borland, und Dado Banatao from the Philippines (S3, Chips and Technologies). For the period 1995-1998, 29 percent of the high-tech startups in Silicon Valley, a total of some 1200 companies, were run by Indian or Chinese immigrants and accounted for almost $17 billion in sales and almost 60,000 jobs (Saxenian, 2000: p. 253). To this corresponds the pattern of admission and graduation in the leading engineering schools of the region and beyond: in the U.S. as a whole, the number of doctorates in science and engineering granted annually to immigrants from China and India has more than tripled and doubled, respectively, between 1990 and 1996, accounting, together with Taiwanese, for 62 percent of all foreign doctorates in science and engineering. At California's universities, the number of Asian doctorates in science and engineering is again twice what it is in the U.S. at large (Saxenian, 2000, p. 250). These figures indicate the extraordinary extent to which the success of Silicon Valley depends on foreign talent; they also explain the development of a flourishing network of technical and commercial relationships between Silicon Valley companies and the home regions of these immigrants.

16.5 Affinity Matters: Higher Education and the Culture of Change

I have already referred to the importance that authors like John Seely Brown attach to the principle of reciprocity where the creation and utilisation of knowledge is concerned, and to the many ways in which the relationship between higher education and ICT in Silicon Valley manifests this principle

particularly well. This is because of the 'close interaction and mutual exchanges of the people involved' that physical proximity makes possible, what The Economist called 'F2F', face-to-face (*The Economist*, 2002: p. 50). Proximity, however, is not the whole story.

The other reason why the relationship between higher education and the development of the ICT industry in Silicon Valley has developed into such a truly symbiotic partnership is the cultural affinity between these partners, the degree to which the universities involved resonate to the particular cultural traits of their corporate partner institutions, and vice versa. There is not, as there is in many other parts of the world, a 'cultural divide' between the world of high-tech business and the world of higher education.

In reviewing this particular and rather striking kind of compatibility and correspondence, I am very much aware of two caveats: First, that this kind of affinity may be not only an asset, but a liability as well. Here as in other respects, it is important to listen carefully to the critical voices that I have already cited (see also Kirp and VanAntwerpen, 2002). Secondly, even this rather remarkable degree of compatibility has its limitations; as an economist who freely admits to 'love the market', William Bowen considers it necessary, in his remarkable Romanes Lecture at Oxford two years ago, to emphasise 'that universities are not businesses (though they have many business-like aspects)' (Bowen, 2000: p. 3).

With these reservations in mind, I am going to review some of the elements of this correspondence that make for such ready and remarkable reciprocity between the academic world of Northern California and the high-tech industry of Silicon Valley.

a   One of the very basic elements of affinity between institutions of higher education in the United States and the corporate world is that both subscribe to an entrepreneurial paradigm of operation. That is not unusual for companies, but it is relatively rare for universities outside of the U.S., even though Burton Clark, when he went out to find some that had emulated the American 'entrepreneurial university' model claimed that he found a few (Clark, 1998). Clearly, the cultural match even in the U.S. is limited: universities do march to a different tune when compared to commercial enterprises, but their willingness to explore new ventures, to take a certain amount of risk, and to remain flexible in placing resources behind new goals make them generally more open to the kind of cooperation that has emerged between, for example, Stanford and the companies of the Stanford Industrial Park.

b   American universities, especially the better ones, appear to have a particularly robust sense of their own independence and self-determination. I remember a conversation between Mr. Biedenkopf, then the Prime Minister of Saxony, and the president of Stanford University, John Hennessy, in which Biedenkopf was wondering whether Stanford was not afraid of undue outside influence considering the large amount of outside money it received, much of it from corporate sources. Hennessy's answer was that, on the one hand, Stanford would never think of accepting any money that had any strings attached, and that, on the other hand, donors dealing with Stanford would know better than to expect buying influence over Stanford's research and teaching agenda. Now that is clearly a bit too good to be true, and American higher education is not quite that immune to seduction (Noble, 2001; Aronowitz, 2000), but generally the cooperation between academia and Silicon Valley companies has benefited from a mutual respect for the independence of the partner, and from a rather relaxed and self-assured attitude on the part of universities vis-à-vis the corporate world. Traditionally, European universities have been a great deal more nervous about these contacts.

287

c From the point of view of the university industry relationship and its success, one of the key ingredients in the institutional make-up of American universities is the rather unique construct of the 'professional school'. Different from academic departments like Psychology or Economics or Political Science, professional Schools of Engineering, Schools of Business, Schools of Law, or Schools of Education form bridges between the world of academic research and the world of professional practice, committed and held to the rigorous academic standards of the institution, but at the same time deliberately open to the knowledge needs of the worlds of technology, of business, of legal affairs, or of education. They tend to be interdisciplinary and structured around areas of professional concentration, and serve as a particularly congenial vehicle for the kind of interaction that has emerged between places like Stanford and Silicon Valley. It is not surprising that the key institutional players in this relationship were, on the side of the universities, Schools of Engineering and Business, drawing their strength and expertise from a wide range of disciplines throughout the university.

d Another characteristic of American higher education that has served particularly well in the kind of relationship we are considering here is its differentiated nature, i.e. the fact that it covers a broad spectrum of very different kinds of institutions of higher education along a gradient of higher or lower selectivity, from highly selective private and public institutions like Stanford and Berkeley to the community-based junior colleges such as, in the case of the Silicon Valley neighbourhood, Foothill College or de Anza College. This kind of differentiation allows the system to optimise the accommodation of students with a wide variety of interests, aspirations and talents; it also provides a calibrated range of cooperative possibilities for the relationship between higher education and the high-tech industry, from cooperation in cutting-edge technology research in institutions like Stanford's Center for Integrated Systems (CIS) to the very successful training programmes for engineering and technical staff in which companies like IBM and SUN cooperate with the local junior colleges. The kind of focus that this division of labour allows each type of institution would be very difficult to achieve in the kind of all-purpose university that is more common in Europe.

e Just as in the corporate world of Silicon Valley, it is true of higher education in the United States that personal leadership matters. It matters in different ways, given the different nature of the institutions, but just as it is hard to write the history of Silicon Valley without paying tribute to the likes of David Packard, William Hewlett, Robert Noyce (Fairchild – Intel),

Gordon Moore (Fairchild – Intel), Steve Jobs (Apple) and the many others who shaped that unique corporate culture, it is difficult to overestimate the role that outstanding academic leaders like Frederick Terman, William Miller and James Gibbons at Stanford or David Kerr and Richard Atkinson at the University of California have played in making their institutions the kind of high-quality centres of scientific excellence that would both inspire, and benefit from, the kind of technological breakthroughs that mark the history of Silicon Valley. Not the least important part of that academic leadership, incidentally, was the attention it paid to the balanced intellectual growth of their universities; while it would have been easy to give in fully to the lure and luxuries of technology, both Stanford and Berkeley stand out nationally and internationally for having maintained a sound and solid base in the humanities and the social sciences as well as in the natural sciences and engineering.

f It is this broad-based intellectual competence at universities like Stanford and Berkeley that is also capable of nurturing the kind of critical discourse on the relationship between education and information and communication technology that has become the more important, the more dominant a social force that relationship has become. That critical discourse stems from different intellectual traditions and has by now engaged a number of scholars and thinkers within and outside the United States (Aronowitz, 2000: Noble, 2001; Press & Washburn, 2000), but it is no accident that some of the most serious critical reflections on the relationship between technology and education come out of the very universities that figure so prominently in the growth of Silicon Valley: from Larry Cuban's hard-nosed assessment ('Oversold and Underused' Cuban, 2001) of the role of computers in the classroom to Hubert Dreyfus' philosophical treatment of 'the limitations of life in cyberspace' (Dreyfus, 2001). And Nicholas Burbules, who with Thomas Callister has become one of the most penetrating analysts of 'the promise and the challenge of new technologies' for higher education (Burbules, 2000a, 2000b), has a PhD from Stanford.

g I have already discussed what an important role the talents of immigrants have played in the Silicon Valley story. Here again, the openness of the industry has been matched and reinforced by the degree to which graduate programmes in science and engineering have been opened up to foreign graduate students, with California institutions being much more open than the national average, especially where students from Asia are concerned. This policy has produced a steady and increasing stream of well-trained scientists and engineers for the companies of Silicon Valley, which in turn has produced its own generation of high-tech immigrant entrepreneurs, many of them with excellent and mutually beneficial relationships with their home countries. Some serious doubts are now being cast upon this policy in the light of some of the United States government's post-9/11 measures; Silicon Valley would be particularly affected by any curtailments and restrictions in the access of foreign technological talent to American graduate training.

h   Lastly, one of the linkages between universities and high-tech industry that is not to be underestimated is the trend in American higher education towards the steadily increased use of ICT in its own academic operations. This trend is well documented for the country as a whole (Harley et al., 2002), but, not surprisingly, it is particularly pronounced in Northern California. Stanford is, once again, a case in point, with its Center for Professional Development (DiPaolo, 2002), its partnership in the UNext and the 'Alliance for Lifelong Learning (AllLearn)' on-line and distance education ventures, and the major initiative, with the help of the Wallenberg Foundation, of the Stanford Center for Innovations in Learning (SCIL) for systematically exploring and advancing the educational use of technology (http://scil.stanford.edu).

290

16.6 The Limits of Replication: Higher Education and IT Entrepreneurship in Europe

The general wisdom in the literature regarding the chances of replicating elsewhere the peculiar symbiosis between higher education and high-tech industry of Silicon Valley is decidedly sceptical. Brown and Duguid have developed an 'ecological model' of what has happened in Silicon Valley, an ecology that 'is built ... through shared practice, face-to-face contacts, reciprocity, and swift trust, all generated within networks of practice and communities of practice' (Brown & Duguid, 2000: p.37). This ecology, they argue, is very hard to replicate, even at enormous cost (for such things as the Valley's leading universities and other infrastructure, for example) because 'knowledge ecosystems develop over time, building connections between participants until they reach a critical mass and take on a collective dynamic all their own.' (Brown & Duguid, 2000: p.38)

There is much to be said for concurring in this cautious view of replication. To make both proximity and affinity work has taken many years even where, as in Silicon Valley, both were present to a remarkable degree.

By and large, the situation in other parts of the world is characterised by limited proximity, even more limited affinity between higher education and high-tech development, and a great deal less of the kind of 'shared experience' that Brown and Duguid consider such an important prerequisite for the functioning of the 'ecological model'.

Let me elaborate a little with reference to the country that, outside of the United States, I know best: Germany. I suspect, however, that the situation is not dramatically different in the Netherlands or elsewhere in Europe.

There is no question but that the ICT industry in Germany and in other European countries has developed significantly in recent years, and certain zones of concentration have emerged, notably around Munich, Dresden, and Berlin. There has also been a significant development in the creation of 'science and technology parks' all over Europe, many, though by no means all, of them linked to one or more institutions of higher education, such as the one in Aachen, the Barcelona Science Park, Sophia-Antipolis in Southern France, the Centuria Science and Technology Park near Bologna, Silicon Fen in Cambridge with links to Trinity College and the ambitious project in Adlershof in Berlin that is closely connected to the Humboldt University (Galbraith, 2002).

291

Furthermore, and as the chapters in this book amply document, major efforts have been undertaken to open up European universities to the possibilities of ICT both as a field of research and development and as an instrument for more effective teaching and continuing education (cf. also Commission of the European Communities 2001); the OECD classifies this latter development as 'high' in the UK and Scandinavia, as 'medium' in Germany and France, and as 'low' in Southern Europe (Larsen, 2002: p. 77). The German Federal Ministry of Education and Research has been particularly active in supporting IT developments in higher education such as the 'notebook universities', and the joint federal-state commission for educational planning (Bund-Länder-Kommission für Bildungsplanung und Forschungsförderung, BLK) has played an important role in coordinating federal and state initiatives in this realm (BLK 2002; Kleimann & Berben, 2002). Institutional and inter-institutional efforts range from the established traditions of European distance education at institutions like the FernUniversität Hagen and the 'Open University of the Netherlands' (Curran 2002) to more recent ventures such as the 'Virtueller Campus II', a joint effort by the universities of Hanover, Hildesheim and Osnabrück to more effectively integrate mult-imedia forms of teaching (CHE 2002: p.11).

However, all of this considerable advancement in the linkages between higher education and ICT notwithstanding, I am arguing that the kind of symbiotic relationship that has been at the very core of the Silicon Valley experience remains an elusive goal in settings such as Germany – at least until and unless some major changes take place.

To back up my thesis, I go back to the analysis of what has made Silicon Valley, even by American standards, such a unique habitat for a symbiosis of higher education and technological development and innovation.

On the face of it, proximity would seem to be the one ingredient in the Silicon Valley formula that it would be easiest to emulate in Europe, particularly in a setting that, compared to the wide open spaces of the American West, is pretty condensed to begin with. However, proximity, as we have seen in Silicon Valley, is not just geographical closeness; if it were only that, regions such as the greater Munich area or Dresden would be good candidates, just as Route 128 in Massachusetts should have worked much better than it did. But to physical presence needs to be added the willingness and the ability to make use of its opportunities, the capacity to engage in 'F2F', in face-to-face interaction on a sustained basis, to generate what Brown and Duguid, using Alfred Marshall's famous expression, have called 'the mysteries in the air' of Silicon Valley, the shared community of discourse, knowledge, practice, and trust (Brown & Duguid,2000: p. 20ff.). It is this shared community that is much harder to come by in a setting such as Germany where the two cultures of academia and entrepreneurship still have very little in common.

By way of elaborating on this basic proposition, let me point out a number of more specific difficulties that would lead me to counsel caution against any easy hopes to replicate the Silicon Valley experience elsewhere, but that one would also have to bear in mind in any attempt to move in the direction of closer interaction between these two cultures.

a   Cultures of entrepreneurship have, as we have seen, their own traditions and value systems. This is why they are not easily transplanted and copied. At the same time, certain traits in corporate cultures are obviously more conducive than others to innovation and change and, especially in knowledge-based industries, to interaction with higher education. It is here that, for the corporate culture of Germany, one would have to note distinct deficits in the acceptance of experimentation and risk, in the tolerance of failure, in the value that is attached to achievement and, quite notably, in the willingness to bring in talent from all over the world.

b Similarly, any honest assessment of the academic culture in European universities would reveal a number of traits that are not, to put it mildly, easily compatible with an entrepreneurial culture of innovation and change. Those traits would include relatively rigid and inflexible organisational and decision-making structures, a relatively weak tradition of cooperative research, especially of an interdisciplinary nature, a relatively under-developed relationship to professional practice (of the kind that has found a legitimate institutional form in the American professional school), and a traditional image of the professoriate as a self-sufficient entity with which the notion of academic entrepreneurship is not easily compatible. There has also been, with some notable exceptions, a reluctance in European universities, reinforced by rather restrictive admissions regulations, to open their doors too wide for talented graduate students from other parts of the world. Much of this, I should hasten to add, has begun to move and change over the last ten years, and there are exceptions to the summary statements expressed here, especially in institutions like the German Fachhochschulen ('Universities of Applied Science') which have taken a much less inhibited interest in open cooperative relationships with industry and business. On the whole, however, the overall tendency in that part of European higher education that I know something about is still one of considerable reluctance to get involved in a steady and institutional interaction with the corporate world.

293

c This reluctance becomes quite apparent in the actual state of the relationship between academia and the corporate sector in countries like Germany. There is very little of the 'trust' that Brown and Duguid consider so essential for a functioning 'knowledge ecology'; instead, there still is, on both sides, a strong residual element of suspicion of the other's motivations and values. This is accompanied, and no doubt reinforced, by a good deal of ignorance about one another an ignorance that is only gradually giving way to better mutual understanding through such devices as corporate membership on university boards, joint programmes of continuing education, cooperative research projects and the like. Ever so gingerly, steps are being taken in recent German legislation on faculty remuneration to facilitate the migration into and out of academia of scholars who see their natural home in both the university and the corporate world. This is still a long way from the ease with which Stanford professors move between the Stanford School of Engineering and their Silicon Valley start-up companies down the road, and there may be some virtue in not making this process too easy. Without a further erosion of the boundaries between higher education and the corporate world, however, neither will the innovative benefits of closer cooperation be reaped nor the important possibilities for critically assessing the work of the corporate sector be realised.

16.7 Conclusion

The key argument of the earlier parts of this chapter was that it takes the combination of regional proximity and cultural affinity between higher education and the ICT industry to explain the Silicon Valley experience. This argument holds in reverse as well when it comes to understanding why it might be difficult in other parts of the world (or, for that matter, the United States) to replicate this experience simply by putting universities and high-tech companies together in the same area. 'Aspiring regions can clearly learn from established ones', concede Brown and Duguid, but the kind of 'knowledge ecosystem' that they have identified in Silicon Valley does seem to defy mechanical imitation (Brown & Duguid, 2000: p. 38).

References

Aronowitz, Stanley. (2000). *The Knowledge Factory: Dismantling the Corporate University and Creating True Higher Learning*. Boston: Beacon Press.

Bowen, William G. (2000). *At a Slight Angle to the Universe: The University in a Digitized, Commercialized Age* (Text of the Romanes Lecture, Oxford University, October 17, 2000). (Unpublished manuscript).

Brown, John Seely, & Duguid, Paul. (2000). Mysteries of the Region: Knowledge Dynamics in Silicon Valley. In: Chong-Moon Lee, William F. Miller, Marguerite Gong Hancock, & Henry S. Rowen (eds.). *The Silicon Valley Edge: A Habitat for Innovation and Entrepreneurship*, pp. 16-39. Stanford, ca: Stanford University Press.

Bund-Länder-Kommission für Bildungsplanung und Forschungsförderung (BLK). (2002). *Strategiepapier Breiter Einsatz von Neuen Medien*. Bonn: BLK.

Burbules, Nicholas C., & Callister, Thomas A. (2000a). *Watch It: The Risks and Promises of Information Technologies for Education*. Boulder, CO: Westview Press.

Burbules, Nicholas C., & Callister, Thomas A. (2000b). Universities in Transition: The promise and the Challenge of New Technologies. *Teachers College Record* 102, 2, 271-293 (http://www.tcrecord.org – ID No. 10362).

Business Week. (1997). Silicon Valley (Special Issue). *Business Week*, August 25, 64-147.

Castells, Manuel. (1996). The Rise of the Network Society. Oxford: Blackwell.

Castilla, Emilio J., Hwang, Hokyu, Granovetter, Ellen & Granovetter, Mark. (2000). Social networks in Silicon Valley. In: Chong-Moon Lee, William F. Miller, Marguerite Gong Hancock, & Henry S. Rowen (eds.), *The Silicon Valley Edge: A Habitat for Innovation and Entrepreneurship*, pp. 218-247. Stanford, CA: Stanford University Press.

Centrum für Hochschulentwicklung (CHE). (2002). Checkup, May.

Clark, Burton R.(1998). *Creating Entrepreneurial Universities: Organizational Pathways of Transformation*. Oxford: Pergamon.

Commission of the European Communities. (2001). *The eLearning Action Plan: Designing tomorrow's education* (Communication from the Commission to the Council and the European Parliament – com(2001)172 final). Brussels: European Commission.

Cuban, Larry (2001). *Oversold and Underused: Computers in the Classroom*. Cambridge, MA: Harvard University Press.

Curran, Chris (2002). Universities and the Challenge of E-Learning: What Lessons from the European Open Universities? In: Diane Harley, Shannon Lawrence, Sandra Ouyang, and Jenny White (eds.), *University teaching as E-Business? Research and Policy Agendas*, pp. 81-87. Berkeley: University of California Center for Studies in Higher Education.

DiPaolo, Andy (2002). On-line Education: The Rise of a New Educational Industry. In: Diane Harley, Shannon Lawrence, Sandra Ouyang, & Jenny White (eds.), *University teaching as E-Business? Research and Policy Agendas*, pp. 61-70. Berkeley: University of California Center for Studies in Higher Education.

Dreyfus, Hubert L. (2001). *On the Internet*. London: Routledge.

Galbraith, Kate. (2002). Technology Parks Become a Force in Europe. *The Chronicle of Higher Education*, January 11. (http://chronicle.com/weekly/v48/i18/18a05001.htm)

Gibbons, James F. (2000). The Role of Stanford University: A Dean's Reflections. In: Chong-Moon Lee, William F. Miller, Marguerite Gong Hancock, and Henry S. Rowen (eds.), *The Silicon Valley Edge: A Habitat for Innovation and Entrepreneurship*, pp. 200-217. Stanford, ca: Stanford University Press.

Grindley Peter C., & Teece, David J. (1997). Managing Intellectual Capital: Licensing and Cross-Licensing in Semiconductors and Electronics. *California Management Review* 39, 2, 8-58

Harley Diane, Lawrence, Shannon, Ouyang, Sandra & White, Jenny (eds.). (2002). *University teaching as E-Business? Research and Policy Agendas*. Berkeley: University of California Center for Studies in Higher Education.

Henderson, Rebecca, Jaffe, Adam B. & Trajtenberg, Manuel. (1998). Universities as a Source of Commercial Technology. *Review of Economics and Statistics*, 80, 1, 119-127

Hirsch, Paul M. (1972). Processing Fads and Fashions: An Organization-Set Analysis of Cultural Industry Systems. *American Journal of Sociology*, 77, 639-659

Kaplan, David A. (1999). *The Silicon Boys and their Valley of Dreams.* New York: HarperCollins.

Kirp David L., & VanAntwerpen, Jonathan. (2002). Academic E-Collaborations and Old-School Rivalries. *The Chronicle of Higher Education,* June 28.

Kleimann Bernd, & Berben, Tobias. (2002). *Neue Medien im Hochschulbereich: Eine Situationsskizze zur Lage in den Bundesländern* (HIS-*Kurzinformation* B3/2002). Hannover: HIS, 2002 (http://www.his.de/doku/presse/pm/pm-kibo302.htm)

Larsen, (Kurt). 2002. Internationalization of Higher Education: oecd Perspectives. In: Diane Harley, Shannon Lawrence, Sandra Ouyang, & Jenny White (eds.), *University teaching as E-Business? Research and Policy Agendas,* pp. 75-80. Berkeley: University of California Center for Studies in Higher Education.

Lee, Chong-Moon, Miller, William F., Hancock, Marguerite Gong & Rowen, Henry S. (eds.). (2000). *The Silicon Valley Edge: A Habitat for Innovation and Entrepreneurship.* Stanford, CA: Stanford University Press.

Lee, Chong-Moon, Miller, William F., Hancock, Marguerite Gong & Rowen, Henry S. (2000). The Silicon Valley Habitat. In: Chong-Moon Lee, William F. Miller, Marguerite Gong Hancock, & Henry S. Rowen (eds.), *The Silicon Valley Edge: A Habitat for Innovation and Entrepreneurship,* pp. 1-15. Stanford, ca: Stanford University Press.

Noble, David F. (2001). *Digital Diploma Mills: The Automation of Higher Education.* New York: Monthly Review Press.

Press, Eyal, & Washburn, Jennifer. (2000). The Kept University. *Atlantic Monthly,* 285, 3 (March 2000), 39-54.

Rowen, Henry S. (2000). Serendipity or Strategy: How Technology and Markets Came to Favor Silicon Valley. In: Chong-Moon Lee, William F. Miller, Marguerite Gong Hancock, & Henry S. Rowen (eds.), *The Silicon Valley Edge: A Habitat for Innovation and Entrepreneurship,* pp.184-199. Stanford, ca: Stanford University Press.

Saxenian, Annalee. (1996). *Regional Advantage: Culture and Competition in Silicon Valley and Route 128* (2nd edition). Cambridge, MA: Harvard University Press.

Saxenian, Annalee. (2000). Networks of Immigrant Entrepreneurs. In: Chong-Moon Lee, William F. Miller, Marguerite Gong Hancock, & Henry S. Rowen (eds.), *The Silicon Valley Edge: A Habitat for Innovation and Entrepreneurship,* pp. 248-268. Stanford, CA: Stanford University Press.

The Chronicle of Higher Education. (2002). Brains and Bucks: How Colleges Get More Bang (or Less) from Technology Transfer. *The Chronicle of Higher Education,* July 19.

The Economist. (1997). A Survey of Silicon Valley (Special Section). *The Economist,* March 29, following p. 60.

The Economist. (2002). Face-to-face communications: Press the flesh, not the keyboard. *The Economist,* August 24, 50-51

The Next Silicon Valley Leadership Group (ed.). (2001). *Next Silicon Valley: Riding the Waves of Innovation (White Paper, December 2001)*. San Jose, CA: Joint Venture: Silicon Valley Network. (www.jointventure.org)

Weiler, Hans N. (1998). Hochschulen und Industrie in den usa – Lehren aus Silicon Valley. Perspektive – *Brandenburgische Hefte für Wissenschaft und Politik*, Heft 5 (Sommer 1998), 53-66. (also: Vereinigung der Unternehmensverbände in Berlin und Brandenburg (uvb) (ed.), *Wissenschaft und Wirtschaft*. Berlin: uvb, 1998, 9-18)

Weiler, Stephan, Thompson, Eric & Ozawa, Terutomo. (2001). The Evolution of a New Industrial District: The Automobile Industry in the American Southeast. *Planning and Markets*, 4, 1, 24-29

About the Authors

Dr. Anne Auban is an associate professor, and Director of the Multimedia Production Unit, University Pierre & Marie Curie Paris, the largest French scientific university. She is commissioned by her president to foster international development in ODL. She is a European expert to the European Community for telematics and multi-media applications programs for education and vocational training: Delta, Telematics, IST, TEN... Dr. Auban is also Vice President of the International Council for Open & Distance Education (ICDE).

Drs. Wim de Boer is one of the educational designers working on the TeleTOP team of the Faculty of Behavioural Sciences, and also a member of the Department ISM where he is doing his PhD research on models of e-learning for flexibility and enrichment.
Research interests: Implementation of technology in an educational institution, design of www-based course-support systems, adaptability of www-based course-support systems. For an overview of his work and background, see http://users.edte.utwente.nl/boerwf.

Drs. Petra Boezerooy graduated in Public Administration at the University of Twente in 1997 and started as a research associate at CHEPS in the same year. Since 1997, one of her research activities relates to the CHEPS Higher Education Monitor (CHEM) which is an ongoing research project commissioned by the Dutch Ministry of Education, Culture and Science. CHEM consists of a number of different outputs, providing in-depth and up-to-date comparative insights into the higher education infrastructure and policies in nine Western European countries: Austria, Belgium (Flanders), Denmark, Finland, France, Germany, the Netherlands, Sweden and the United Kingdom. Since December 1999, she has been involved in various (mainly comparative) projects concerning the use of information and communication technologies (ICT) and distance education in different countries. Right now, she is working on a one-year survey called 'Future scenarios for the use of ICT in higher education: an international comparative study' in which the focus is on which scenarios are emerging with respect to the use of ICT in higher education and how future developments can be predicted and strategic choices be based on that? Since January 2001, Petra has been the academic officer of the postgraduate Distance Learning programme on 'Institutional management and change in higher education'. Furthermore, as a result of the above projects, Petra has coordinated and organised several study visits (for both the Dutch Ministry of Education, Science and Culture and Stichting SURF) in the last two years. In the year 2000 to both the UK (spring 2000) and the USA (autumn 2000) and last March to Australia. Another study trip to the USA is planned for November 2002.

Jim Devine is Director of the Institute of Art, Design & Technology, at Dun Laoghaire, Dublin. This is a recently established polytechnic higher education institution specialising in media arts, visual arts, enterprise and technology. Prior to joining IADT-DL, Jim Devine was a senior member of staff at OSCAIL, the National Distance Education Centre, based in Dublin City University. He is a member of the EUA expert group on new technologies for teaching and learning in higher education and also participates regularly in EU expert groups and as an evaluator within the 5th Framework IST programme and the eLearning initiative under DG Education & Culture.

299

Dr. Claudio Dondi, born in Modena in 1958 and an industrial economist as a university background, is the President of SCIENTER, a non-profit research organisation based in Bologna and active Europe-wide in the field of innovation of education and training systems, since its establishment in 1988. In this position his main activities are the coordination of large national and European projects, as well as policy advice and evaluation at regional, national and international level. His other positions include: Professor of Human Resource Development at the College of Europe in Bruges, Member of the Board of the MENON EEIG in Brussels, Member of the Editorial Boards of the British Journal of Educational Technology and of the European Journal of ODL, Vice-President of the European Institute for e-learning, Member of the Executive Board of EDEN, European Distance Education Network, and Member of TTnet Italy.

Dr. Peter Floor (born 1934) studied geology at Leiden University, Netherlands. After having been a staff member of the Geology Department he formed part of the Leiden University administration with responsibility for educational and research policy issues from 1972 till 1997. He became involved in the Coimbra Group of Universities (created in 1985; www.coimbra-group.be) in 1986 and acted as chairman of its Steering Committee till early retirement in 1997. Now retired, he remains involved in some Coimbra Group affairs giving special attention to strategic processes of universities regarding overall implementation of new technologies in their teaching.

Maruja Gutiérrez Díaz works at the Directorate General Education Culture of the European Commission, where she is head of the unit Multimedia: culture – education – training. For more information, see: http://europa.eu.int/comm/elearning.

Dr. Nick Hammond is a Reader in Psychology at the University of York, UK, and Director of the Learning and Teaching Support Network Subject Centre in Psychology (LTSN Psychology). He is President of the Association for Learning Technology, a UK-based educational organisation which seeks to bring together all those with an interest in the use of learning technology in higher and further education, and editor of the journal Psychology Learning and Teaching. Current research interests cover a range of issues in learning, teaching and assessment in HE, including the use of ICT to support group-based learning.

Dr. Pekka Kess is the Director of The Finnish Virtual University, a project of all universities in Finland to establish a networked service system to support academic education, research and administration. Before becoming its director, he was a project manager of the Finnish Virtual University in charge of the Portal Development project. Pekka Kess holds a position as a Professor of Industrial Engineering and Management at the University of Oulu (currently on leave). He was involved in teaching and research of industrial enterprises and especially their product and technology strategies, core competencies, business process improvement and re-engineering. He was from 1996-1998 the Director of the Centre for Continuing Education of this university. Special interest areas are in strategic management, risk management, new ventures, TQM, global business issues and intellectual capital of organisations. He has been involved in the management of several research projects: HILO (High Tech Logistics), ELVA (TQM of Electronic Manufacturing Networks), NIPPU (Quality Improvement in contract manufacturing networks in electronics, timber and food industries), ProElectronica (Electronics manufacturing improvement) and study on its effects on the society in the Oulu Region, eLISE (Creation of a global technical service and maintenance concept for the electronics and telecommunication industries). Professor Kess has been involved in a wide range of European projects both in the field of engineering and in the field of continuing education and lifelong learning. He was also involved in work for the Finnish Higher Education Evaluation Council in evaluating Tampere University (1999-2000) and the Open Universities of Finland (2001-2002).

Dr. Bernard Levrat holds a PhD in physics (Ann Arbor, Michigan, USA, 1963) and is a Professor of Computer Science at the University of Geneva. He previously held the position of Vice-Rector of his university (1991-1995) and of Director of RERO (network of University libraries for the French-speaking part of Switzerland) (1996-2001). Since 1996 he has been the Chairman of the Working Group of the Swiss University Conference on Learning and NTIC which became the Swiss Virtual Campus program in 2000.

Dr. Bernard Loing is a university professor (English Language, literature, language teaching, linguistics). He was the former Deputy Minister of French Post and Telecommunications (1981-85). Former Rector of CNED (the National Center for Distance Education in France) (1990-93). He is one of the main French experts in Information and Communication Technology, and its applications to education and training at all levels, with special reference to the field of open and distance learning (ODL). At present, Dr. Loing also serves as the Intergovernmental Liaison Officer and General Delegate of ICDE (International Council for Open and Distance Education) at UNESCO, the President of Canal EF, the digital satellite radio channel for French-speaking Africa, and as the Chairman of the Institute for Information Technology at the Francophone Agency.

301

Dr Peter Maassen is senior researcher at NIFU, Norway, and senior fellow at the Center for Higher Education Policy Studies (CHEPS), University of Twente, the Netherlands. In addition he is the director of Hedda, a consortium of European research centres and institutes in the field of higher education research. He specializes in the public governance (including policy reform and institutional change) of higher education. In 1998 he was awarded a fellowship by the Society for Research on Higher Education (SRHE).

Robin Middlehurst is Director of Development and Professor of Higher Education at the University of Surrey, England. Middlehurst joined the University of Surrey as a research assistant in late 1996 and after 4 years, she joined the Institute of Education, University of London as a lecturer. She joined the Higher Education Quality Council first as an Assistant Director and then as

Director for Quality Enhancement. She led the Group in the design and implementation of a number of developmental programmes including external examining, modularity, managing for quality and the national 'Graduate Standards Programme'. Up to August 2000, she had overall responsibility for management of the Centre for Continuing Education, including the teaching programmes within Work-based learning, the Combined Studies degrees and Open Studies. Prior to entering higher education, she had taught at all levels of education (primary, secondary, further and adult), having gained a PGCE in 1980 and subsequently having undertaken research on social education to gain an MPhil in 1985.

Research on higher education in three main areas inform teaching: policy and practice on quality and standards; leadership and management theory and practice; and most recently, developments in corporate and e-learning and their implications for higher education. Her teaching activities include among other: Leadership and management development programmes for several universities in the United Kingdom. For more information, see: http://www.surrey.ac.uk/Education/profiles/middleh.htm.

Dr. Monika Lütke-Entrup is a project officer at the media division of the Bertelsmann Foundation, Gütersloh, managing a national ICT staff development project. Her areas of interest are lifelong learning, e-learning in higher education and in professional training as well as structural reform and strategic development of German education institutions. Prior to her current position, she worked for the strategy practice of the consulting firm Accenture. She advised clients on ASP/BSP business models and customer relationship management. She also conducted a study on investor relations and shareholder value. Monika Lütke-Entrup holds a D.Phil. from the University of Oxford (UK) and a Teacher´s Degree from the University of Münster (Germany). She has carried out research for the Organisation of American States (Washington, DC) and taught at the Colegio de México (Mexico).

Dr. Bjørn Stensaker is a political scientist, working at the Norwegian Institute for Studies in Research and Higher Education (NIFU) in Oslo as a senior researcher. His research interests are related to studies of how the concept of quality is adapted in higher education, organisational change and studies of institutional management and management systems. He is the author and editor of a number of books and reports on evaluation and quality improvement in higher education (in Norwegian), and has conducted studies on the effects of external evaluation systems in all the Scandinavian countries. Stensaker has also published articles on this topic in a number of journals including *Higher Education, Higher Education Quarterly, Higher Education Policy, Quality in Higher Education, Tertiary Education and Management* and *Journal of Higher Education Policy and Management*. Stensaker is a member of the Executive Committee of the European Association of Institutional Research (EAIR), a Co-Editor of *Tertiary Education and Management* and one of the Executive Editors of *Quality in Higher Education*.

Dr. András Szucs (1956), graduated as a bio-engineer in 1980, After ten years of university teaching at the Budapest University of Technology, from 1990 he held international posts as Director of the EU TEMPUS Programme in Hungary (1990-95), Director of the EU Phare Central-Eastern European Distance Education Programme (1994-96) , Director of the European Communication Strategy Programme of Hungary (1996); since 1997, Executive Director and from 2000 Secretary General of the European Distance Education Network (EDEN). He has been Director of the Distance Education Centre of the Budapest University of Technology and Economics since September 2000.

Dr. Maarten van de Ven is senior consultant and researcher at the Educational Expertise Centre Rotterdam (OECR), part of Erasmus University. He has a Master's degree in Psychology (1983) and a PhD degree from Delft University of Technology (1998). His research and consultancy has been focussed on the use of information and communication technologies in educational processes. Before entering the academic world, he was Production Manager at Courseware Europe bv. For more information see http://www.oecr.nl/Engels/atwork/maarten.html.

Dr. Hans N. Weiler is a Professor Emeritus of Education and Political Science at Stanford University, and a Professor Emeritus of Comparative Politics and past Rektor of Viadrina European University at Frankfurt (Oder), Germany. Currently, he is affiliated with the Stanford Institute for Higher Education Research (SIHER). Among his many activities, he was director of the International Institute for Educational Planning of UNESCO in Paris. See http://www.stanford.edu/~weiler for more information.

Dr. Marijk van der Wende is a professor and senior researcher at the Centre for Higher Education Policy Studies (CHEPS) at the University of Twente. She holds a chair in comparative higher education studies with special reference to the impact of globalisation and network technologies. Van der Wende has a Bachelor degree in teaching (1980), pedagogy (1984) and a Master's (1991) and PhD degree (1996) in Educational Sciences. Her research has over the last ten years been focussed on the processes and policies for internationalisation in higher education, with particular interest for national policies, internationalisation of the curriculum, the implications for quality assurance and the role of information and communication technology (ICT).